PRAISE FOR *IN SEARCH OF MYCOTOPIA*

"*In Search of Mycotopia* is for mycophiles everywhere, from old hands to young enthusiasts, profiling in lively prose the kooky, brilliant, and inspiring folks who are changing the way we understand fungi, nature, and ourselves."

— EUGENIA BONE, author of *Mycophilia* and *Microbia*

"*In Search of Mycotopia* is as wondrous and hopeful as its awe-inspiring subject. Doug Bierend deftly extends the mycelial threads of his curiosity into the many communities that congregate around fungi, from academic researchers to hipster entrepreneurs to Indigenous groups engaged in cultural mycoremediation. The resultant book is a masterpiece of intersectional fungal anthropology that will send you running for the mushroom stand at your nearest farmers market—and may even inspire you to venture forth on some forest forays yourself."

— BEN GOLDFARB, author of *Eager*

"Fungi are nature's alchemists, the circulatory and nervous system of the forest ecosystem, and the providers of nutrition, healing, remediation, and spiritual awakening. In this compelling book, Doug Bierend awakens the myco-nerd in each of us, uplifts the stories of queer, Black, and young citizen scientists, and challenges us to ameliorate our fungal illiteracy. Bierend convinces us that the healing of our planet requires that we remember the fringe and oft-forgotten fungal kingdom."

— LEAH PENNIMAN, cofounder of Soul Fire Farm,
author of *Farming While Black*

"What if our world were connected by unseen strands, by dynamic webs of life that maintain nature as we know it, a largely hidden substrate capable of healing us, feeding us, teaching us, churning death into life, sustaining the soil, plants, and animals? Well, it is, actually. Amiable, brilliant, and endlessly curious, Doug Bierend is the perfect guide to both the marvelous realm of fungi and to the radical human subcultures that have sprung up to celebrate it—citizen mycologists, cultivators, and activists from the Ecuadorian Amazon to the Pacific Northwest. *In Search of Mycotopia* is a fascinating, humble, and hopeful book, a glimpse at a cosmos of which we are not the center; in which everything is interconnected, and life humbly, quietly persists."

— BEN EHRENREICH, author of
Desert Notebooks: A Road Map for the End of Time

"There's a fungus among us, and it's all good. As Doug Bierend's engaging journey through mycoculture reveals, it's time to make like a mushroom, join the club, and grow a distributed, curious, and sustainably prosperous world together—from the bottom up."

—DOUGLAS RUSHKOFF, host and author of *Team Human*

"Doug Bierend's book is a vivid trip past the flora and fauna of this world and into the reigning kingdom of Fungi. The most mind-altering forays take readers far beyond psychedelic 'shroomer culture, which turns out to be one in a collection of subversive subcultures of enthusiasts, scientists, and citizen scientists from all walks of life. Written by a playful and candid storyteller, *In Search of Mycotopia* vindicates the mushroom, the literal and symbolic fabric that tightly binds together all life."

—CAREN COOPER, author of *Citizen Science*

"*In Search of Mycotopia* is a must-read! Leaving no stone unturned, Bierend covers everything from cultivation to psychedelic studies, providing insight into how critical fungi are toward healing the planet and its inhabitants, and ultimately encouraging readers to become a part of the community. Nothing is impossible if you bring mushrooms into your life, and reading this book is a great way to begin your journey."

—TRADD COTTER, author of
Organic Mushroom Farming and Mycoremediation

"With a fresh and welcoming voice, Doug Bierend takes us on a journey through the far-reaching branches of the modern mycocultural movement. As his search for mycological knowledge expands across the pages, so do we also find ourselves enmeshed in the underground world of citizen scientists, DIY mushroom growers, and landscape healers. Bierend presents voices that are rarely heard elsewhere, and rarer still is his holistic approach to such a diverse and dynamic community. This book offers a rich glimpse into a rapidly growing culture, as well as inspiring insights on the many ways that you can get involved in the future of fungi. A much-needed compass for the ever-evolving landscape of mycology, and a vivid portrait of fungi and the humans who work with them, *In Search of Mycotopia* is a must-read for anyone wanting to join in the fungi fun."

—PETER MCCOY, author of *Radical Mycology*

In Search of MYCOTOPIA

Citizen Science, Fungi Fanatics, and the Untapped Potential of Mushrooms

Doug Bierend

Chelsea Green Publishing
White River Junction, Vermont
London, UK

Project Manager: Alexander Bullett
Editor: Michael Metivier
Copy Editor: Eliani Torres
Proofreader: Diane Durrett
Indexer: Shana Milkie
Designer: Melissa Jacobson
Page Layout: Abrah Griggs

Printed in Canada.
First printing February 2021.
10 9 8 7 6 5 4 3 2 1   21 22 23 24 25

Our Commitment to Green Publishing

Chelsea Green sees publishing as a tool for cultural change and ecological stewardship. We strive to align our book manufacturing practices with our editorial mission and to reduce the impact of our business enterprise in the environment. We print our books and catalogs on chlorine-free recycled paper, using vegetable-based inks whenever possible. This book may cost slightly more because it was printed on paper that contains recycled fiber, and we hope you'll agree that it's worth it. *In Search of Mycotopia* was printed on paper supplied by Marquis that is made of recycled materials and other controlled sources.

Library of Congress Cataloging-in-Publication Data
Names: Bierend, Doug, author.
Title: In search of mycotopia : citizen science, fungi fanatics, and the untapped potential of
    mushrooms / Doug Bierend.
Description: White River Junction, Vermont : Chelsea Green Publishing, [2021] | Includes
    bibliographical references and index.
Identifiers: LCCN 2020047987 (print) | LCCN 2020047988 (ebook) | ISBN 9781603589796
    (hardcover) | ISBN 9781603589802 (ebook)
Subjects: LCSH: Mycology. | Mushrooms. | Fungi.
Classification: LCC QK603 .B453 2021 (print) | LCC QK603 (ebook) | DDC 579.5—dc23
LC record available at https://lccn.loc.gov/2020047987
LC ebook record available at https://lccn.loc.gov/2020047988

Chelsea Green Publishing
85 North Main Street, Suite 120
White River Junction, Vermont USA

Somerset House
London, UK

www.chelseagreen.com

For Vera

If we do not begin to expand our concept of revolution, nature will make the revolution despite us.

—Tawana Petty

Contents

InTRoductioN

W e met Olga Tzogas in front of a featureless warehouse on the southern fringes of Rochester, New York. For three years, the austere building had served as the home of her business, Smugtown Mushrooms, and a haven for her extended community of countercultural misfits. But we were visiting at a complicated moment. In a sudden and grim turn of events a couple of months prior, the landlord had taken their own life. Shocking news, one consequence of which was that the company's fate lay in the hands of a pair of young investors with designs on erecting a hip beer garden in its place.

"These dudes are developing the area and they don't want to renew our lease, so we have to go, and that sucks," said Tzogas, leading my friend Alanna and me into the building. Friendly but flustered in an afternoon rush of activity, Tzogas, a daughter of Greek immigrants and prime specimen of the northeastern anarchomycologist, spoke with equal parts twang and lilt, wearing a tie-dyed tank top, cat-eye spectacles, and the fading tattoo of a forest floor's worth of fungi on her forearm. When she was growing up, Tzogas's family owned a restaurant in Rochester, making for deep and complicated feelings toward the former Kodak boomtown as investor-driven development gradually rendered the local landscape ever less recognizable.[1]

We walked past a cluttered front desk and into a dim, cement-walled room strewn with fungal curios. An ocher pile of reishi caps dried on racks next to a stack of 1970s rock LPs and mushroom-themed books and pins; on a table, rows of mason jars with bespoke labels and cheerily colored cloth tops were stuffed with fungus-inoculated furniture dowels. Crossing the corner and down a step, we entered a dank room—the fruiting chamber, essentially a nursery for mushrooms—with white walls and improvised shelving upholding translucent bags that surged with white mycelia. Pastel-colored clusters of mushrooms burst from the edges of the bags, suffusing the room with a strangely sweet aroma. Plastic tarps stapled to the walls and ceiling helped to prevent contamination and control the balance of oxygen, temperature, and moisture, critical considerations for ensuring the biggest and best-looking mushrooms. All around, a pierced, tattooed, denim-clad crew busily prepared the next run of bags along with jars of liquid culture, a fungal broth that would kick off the transformation of sawdust and grains into delicious mushrooms.

As its founder, Tzogas had been the company's driving force, but to her, turning a profit was secondary to spreading the mushrooms and their message. The motivating mission at Smugtown was to serve as a collaborative node in a wider, nation-spanning network of cultivators, growers, and citizen mycologists, as well as Tzogas's own local community of small-scale farmers, food justice activists, and ecopunks. "We would have parties before everything was built; we would fill up this place with everyone from Rochester and throw down with bands and stuff like that," she said, stepping into a side room to help a workmate juggle beakers in a complex transfer of culture fluids. "Bringing the community together, that's kind of my thing."

As we wound farther into the cavernous facility, I was unknowingly beginning a journey into the heart of the Mycelium Underground. Despite the name, this fungus-focused coalition isn't exactly secretive —it even has a website. Organized by Olga and a handful of her friends around the country, most of them women, it represents one facet of a much wider, intersectional community of amateur mycologists and cultivators, educators and organizers, ecologists and activists "devoted to bringing people together for the love of learning and embracing the world of fungi and all with which fungi [intertwine]."[2] Its goal: to grow mushrooms and teach the potential of these oft-overlooked organisms to serve as food and medicine, partners in environmental remediation, and inspiration for conversations about social justice, the democratization of science, and realizing "a bio-centric world, where we are in tune with ecosystems, and open to all folks of all backgrounds and all walks of life." The focus on mushrooms, in other words, amounts to a lens through which to reassess how we relate to nature and to one another.

Our tour continued, and soon a labyrinthine complex of multicolored corridors led to another, much larger fruiting chamber. On tall racks, row upon row of bags burst with clusters of red reishi, brown shiitake, and egg-white lion's mane. Crowding the floor of the cavernous, graffiti-splattered dock were big blue barrels bearing sawdust and grains from a neighboring brewery. Now used to feed the fungi, these repurposed resources exemplified the capacity mushrooms have for "closing loops" in agricultural and other waste streams. Delicious and valuable mushrooms can grow off spent grains, sawdust, straw, coffee grounds, soybean husks, and more; the list of possible substrates is long. But these upcycling opportunities also benefited Smugtown's bottom line. "One, it's easier," observed Olga. "And two, it's fucking free."

Sequestered in a small side room deep in the heart of the warehouse was the laboratory. Under the yellow light of an incandescent bulb, stacks of petri dishes sat atop repurposed shelves and DIY ducting, while a scavenged HEPA filter and flow hood took up most of the room's compact footprint. Here was Smugtown's inner sanctum, where Olga spent most of her time cloning and cultivating the strains behind her products. All the hundreds of pounds of thriving mushrooms we saw that day had come directly from these tiny, fuzzy splotches on agar plates, calendar dates and strain varieties scrawled on their plastic lids in permanent marker. Despite the slightly bedraggled digs, there were sophisticated methods at work. "I took microbiology at the local community college, so I had the aseptic technique down," boasted Tzogas as we peered into a series of culture plates held up to the light one after the other, fine tendrils of backlit mycelia forming miniature mandalas. "Everyone was fumbling with petri dishes and pouring agar onto plates and stuff and I was just, like, *bing bing bing*."

Next door to the lab was the imposing autoclave, a twenty-foot-long pressurized chamber with a big steel wheel for a hatch. Used for sterilizing high volumes of straw and other substrates to reduce microbial competition for the fungi, it was stuck straight through the wall separating two large rooms, a challenging and expensive installation, the costs of which Olga had unsuccessfully tried to crowdfund. She now faced the same challenge in reverse to have the thing removed. Resembling something between a bank vault and a steamship engine, it added to my growing impression of having stepped aboard a strange voyaging vessel, one whose sailing hadn't exactly been smooth.

"I struggle with getting people that can fulfill the tasks I need," Olga said as we emerged into the garage, where a weathered minivan stood stuffed to the gills with boxes bound for a major mushroom festival in Kennett Square, Pennsylvania, "the Mushroom Capital of the World," where vast farms produce them at commodity scale. The drive there was going to be long, the profit margin slim. "We want to supply mushrooms to the area, regionally," said Tzogas, assessing the minivan's cargo. "Honestly, it'd be cool to get some contract grows for New York and spread out, because Pennsylvania has immense amounts of mushroom farms, and they're supplying eighty percent of the market. And in New York, come on, we can do that."

Underpinning Smugtown's enterprise was a seemingly intractable tension between the demands of running a business and the communitarian, often outright anti-capitalist values Tzogas shares with her community. Their hope is to live and work with a local focus, in accordance with the distributed, integrative dynamics they perceive in nature, and the symbiotic, reciprocal ways of being exemplified by fungi themselves. They flatly reject the logic of extractive capitalism; Smugtown was in business because it had to be, not necessarily because it wanted to be. This tension found clear expression in pointed and often profanity-laden social critique. "Obviously we live in a capitalist system," Tzogas said. "I have to make money to survive, and to pay my stupid gas and electric bills, and pay to get places, and pay people so they can pay their shitty rent and shitty bills, too. Yeah, I totally get it. But I would love to grow mushrooms, maybe sell to some degree, and build my business based on a more mutual aid type of thing, like farmers growing food for other farmers. In my perfect utopian world, we all have these skills that we're sharing, we all help each other survive together.

"I'm not saying I'm not a hustler, let's make that clear," she added after hearing herself speak. "I'm definitely into slinging and getting it done."

As medicinal and culinary markets for mushrooms rise domestically and internationally, the world of mushroom cultivation is growing increasingly competitive, but also more collaborative and distributed. Tzogas's cadre is just one facet of a widespread network of cultivators throughout North America and beyond, often with a local focus and championing principles of seasonal food systems, regenerative land stewardship and food sovereignty, access to natural medicine, as well as values of universal equity and enfranchisement. Accordingly, skills are shared with a priority given to access, such as through the establishing of safe spaces, sliding scale fees for classes and workshops, online courses and videos, books, and various other efforts at lowering the barriers to engagement with fungi and all to which they connect (which, as we'll see, is pretty much everything).

It's hard to think about competition at all when you wander the woods in search of mushrooms. Called a *foray* in its more organized form, this mycological tradition verges on the devotional, representing communion and companionship with fungi in their natural element. Many mycophiles I've met seemingly spend every moment away from the forest thinking about the next chance they'll get to go back. "What am I going to do today?" Olga mused over breakfast on the second day of our visit, bemoaning the mounting challenges facing her business. "Am I going to think about *that* shit, or am I going to go into the woods?"

A couple of hours later, we were strolling under the mottled canopy in the coastal forest along Lake Ontario's southern shore. Superstition among some mushroom hunters says you shouldn't make it too obvious that you're looking for mushrooms, lest they hide. Yet despite our unmistakable intent as Olga, Alanna, and I plodded loudly among the trees, gradually but quite distinctly, the woods around us began to change. Like a pinball machine coming to life, the forest seemed to light up as our foray unfolded, mushrooms appearing all around us, every few steps revealing another new reason to pause, peer, or prod. We came upon the artist's conk, *Ganoderma applanatum*, jutting from the side of a log, with its pale belly that bruises indigo when touched. Halfway up a nearby oak, a bulbous pom-pom of *Hericium erinaceus*, or lion's mane, cascaded from the hole made by a woodpecker. Tiny mycenas—one of the "little brown mushrooms" that we overlook everyday—brought our noses to within centimeters of rotting logs.

Throughout our walk, mushrooms of varying sizes, shapes, and colors seemed suddenly all around, but of course they had been there all along. With Tzogas's experienced guidance, I was just beginning to learn how to see them; I was "getting my eyes," as it's sometimes described. She pulled caps up from the ground in fudgy gobs of dirt, splitting them and lifting the pieces up for a sniff. The distinct and sometimes off-putting aromas of mushrooms are notoriously hard to describe, but they have a special appeal to the fungi cognoscenti, a way of indulging in the sense of abundance that comes from fungal fellowship.

After a few hours we emerged from the forest, our basket filled by a squishy potpourri seasoned with crumbled soil. Back in the van, I was still marveling at how dramatically my picture of the forest had shifted. As we left the parking lot, these musings were interrupted when Olga suddenly swerved to stop on the side of the road, leaping out of the driver's seat and down the grassy embankment. She returned moments later, almost apologetic over the armful of maitake mushrooms that she added to our cargo, having somehow spotted them growing at the base of a tree in the greenbelt that divided the highway.

Back at the warehouse, the Smugtown crew took a break from stocking the minivan, its dashboard littered with dried mushrooms and animal skulls, to peruse our haul. They passed specimens around to one another, turning them over to examine features in the retreating dusk light, handling them reverently.

Alanna and I departed Rochester with bellies full of Chaga tea and sautéed lion's mane, heading back to our shared apartment with a grow bag that would sit on the kitchen counter throughout the winter, lacquered sunset-colored saucers of reishi gradually stretching up and over the edges. Soon trophies from subsequent trips into the woods popped up in various corners of the apartment. Then came the books and the mycological society newsletters, the morphology diagrams and spore prints decorating the walls, the jar of woody Chaga chunks atop the fridge. Eventually, the company of other mycophiles crept into our lives, and signs of our fascination grew in tandem with the depth and breadth of a new network. Alanna received the message of the mushroom loud and clear, promptly becoming enmeshed in the broader myco-community; before long she was growing mushrooms for market and giving talks on female contributions to the fungal arts and sciences. And here I am—quite unexpectedly, I must say—having written a book about the manifold allures and promises of fungi.

Mycophilia spreads as surely as a mycelium itself, and fungi have a knack for persisting in the mind much as they do in a log. They also have a tendency to inspire metaphors and analogies. Some mycophiles joke, or half joke, that their fascination may indeed be traceable to fungal spores having taken root in their brains. That's somewhat unlikely, but what is certain is that the more one discovers about this mysterious kingdom of life, the more it beckons one to further inquiry, appealing to a sense of mystery, possibility, and deepened connection with nature.[3] At least, that's what I tell people when they see the bags of fungus growing in my closet.

Despite their low profile and the ignoble esteem in which many hold them, fungi are largely responsible for life as we know it. In fact, any picture of a healthy, functioning ecosystem is incomplete if not riven with fungi. Their roles as decomposers and nutrient circulators are critical to the formation and regeneration of soils. They form close relationships with the vast majority of leafy plants, often bonding with and weaving within them at the root level to create living networks that give structure to soil and scaffold against erosion, digest pathogens and toxins, and churn carbon into the wider food web. Fungi are constantly devising and secreting enzymes that can break down potential nutrition or fend off microbial competition, and produce compounds that feed, intoxicate, or sometimes kill the animals that eat their reproductive organs, which we call mushrooms.

All these capacities, among many other still more surprising fungal abilities, underpin the emerging field of applied mycology. Mycelia can be trained to consume hydrocarbons or kick-start the trophic cycles that build biodiversity in soils. They are put to work filtering harmful microorganisms from water, such as *E. coli*, and are the basis of a growing field of "mycomaterials" increasingly common in textiles, construction, and shipping. Low-cost mushroom farms can be set up within a matter of weeks, in a wide variety of environments, leading many to see fungi as potential allies in struggles for food security and medicinal sovereignty.

There is promising therapeutic potential in the fact that if you eat the right one—ideally under the right conditions and with the right people—mushrooms may deliver a life-changing shift in consciousness. Eat the wrong ones, of course, and you might get sick or even die, though on the whole there are far fewer deadly mushrooms than one might think; estimates are that about 3 percent of named mushrooms can kill.[4] Nevertheless, parasitic and pathogenic species of fungi can be a scourge to bodies and ecologies alike, a likely reason for the strain of "mycophobia" common throughout North America. Whether helpful, harmful, or benign, one thing is certain: fungi deserve our respect.

Given their ubiquity and utility, it's little wonder fungi have played significant roles in human culture for millennia. Many will think immediately about the mind-altering variety, and the attendant stereotypes of long hair, open-toed sandals, and Grateful Dead T-shirts. This certainly describes a section of North American mycoculture, but age-old spiritual and medicinal traditions in South and Central America, Europe, Africa, and Asia are believed by ethnomycologists to have been influenced by mushrooms' psychoactive qualities. In China, thousands of years of medicinal and culinary mushroom traditions carry on at truly massive scale. In Eastern Europe, foraging for mushrooms is a part of everyday life, such as in Russia, where according to translator Julia Schelkunova, "Calling yourself a mushroom hunter is like calling yourself a pizza eater. You just do it."[5]

In North American culture, our relationship with fungi is comparatively lacking. Indeed, mushrooms are objects of disdain and disgust among many in the Western world. Much of the emerging work around fungi in the United States is about nudging society to be more myco*philic*. The trend lines certainly seem to be pointing in that direction. Mushrooms are quickly gaining popularity, while a broad, diverse, and growing movement is enthusiastically elevating the aesthetic, culinary, medicinal, economic, even heuristic and metaphorical value of fungi. Many devotees meet at increasingly common fungi festivals, convergences, and gatherings popping up around the country. The internet has quite naturally become a vibrant dimension of this mycological renaissance. It serves as the basis of new market opportunities, while hundreds of highly active and growing

Facebook groups, Instagram pages, websites, and other forums reify the network of fungi-fascinated communities spreading around the globe.

In North America, much of the online conversation revolves around foraging, identification, and a thriving DIY cultivation community. People are launching small culinary mushroom businesses that quickly grow into sizable operations, while others make fungi-infused tinctures, coffees, chocolates, and other products; they teach classes and organize community events; they trade strains and cultivation techniques—or *tek*, a term that counts among numerous cultural and practical inheritances from the psychedelic community. Within this expanding circuit of fungus-focused skills sharing and knowledge exchange, a certain subversive strain of subculture is also emerging, one that takes up fungi as part of an effort to dissolve the petrified stumps of patriarchy and colonialism, and establish in their place priorities of biocentrism and universal enfranchisement in science, food, and medicine. That doesn't characterize everybody's involvement—some people, after all, just like mushrooms a whole lot or see a good opportunity to make money—but as I've learned, there exists a real movement to educate people about fungi and, in so doing, draw meaningful social value from them, exemplified by the Mycelium Underground along with other groups with such names as Radical Mycology, Female and Fungi, POC Fungi Community, Fungi for the People, and Decolonize Mycology.

Citizen science is also at the core of these efforts. Despite being a relatively new division within the natural sciences, mycology still embodies many of the same institutional proclivities as the sciences in general. The academic study of fungi, at universities and museums, tends to focus on their genetic diversity and distribution around the globe, how they factor into soil health and ecology, and quite often emphasizing the threats they pose to lumber, agriculture, and other industries. Meanwhile, many basic facts about fungi remain to be uncovered. As of this writing, some 120,000 species of fungi have been identified of an estimated 3.8 million, roughly 5 percent.[6] Meanwhile, besides the expanding market for cultivated mushrooms and their derived products, mycology offers little in the way of professional opportunities. It's little surprise, then, that some of the most interesting and vital happenings in mycology are taking place outside traditional institutions.

Consistent with its marginal status, mycology has always been connected to—and in fact sprang from—communities of dedicated enthusiasts. Universities are increasingly partnering with robust amateur clubs, members of which often have knowledge and expertise that are every bit as comprehensive and refined as that of anyone in academia. Those communities are increasingly diverse, a trend driven largely by greater access to technology and knowledge, enhanced by the culture-making capacity of the internet. There is reason to hope that this dynamic will help shape the future of mycology, and perhaps the natural sciences more broadly, to become more diverse and inclusive.

Beyond science, there is also hope among some that the rising profile of mushrooms can help change the face of society itself, as fungi underscore the immutable interdependence of all living things. Among some traditionally marginalized areas of society, too, there is deliberate work to channel the message of mushrooms in efforts at building solidarity and resilience around food and medicine, exercising agency in the face of oppressive and often racist state power, and even confronting assumptions about gender and other binaries. At the same time, the many new economic opportunities opening up around fungi create pressing questions about who stands to benefit.

Whatever its outcome, the shape and direction of this growing sense of fungal possibilities may be largely determined, like fungi themselves, by its leading edges. The emerging mycological community is directly inspired by these organisms' capacity to renew soils and remediate contaminated landscapes, to provide sustainable sustenance, and, in their very ways of being, to demonstrate how we might operate in more equitable, reciprocal accord with nature—and one another. In other words, fungi may not be fringe, but rather a vibrant front in our collective struggle to realize a better world. Here's hoping.

———————

To be clear, I write this as neither a mycologist nor a scientist. If I am any kind of -ist at all, it is a generalist, and a humble enthusiast when it comes to fungi. But one need not be an expert, however defined, to communicate, innovate, create, and otherwise do beautiful things with or about fungi. The same can be said, I would argue, of anything else in nature. In fact, this book is, in part, about challenging our idea of what counts as expertise, and who gets to participate and make contributions—again, however one chooses to define that. My hope is to uplift a variety of work and ways of relating to fungi, conducted by people of diverse backgrounds, identities, and experience, rooted in respectful partnership with this massive, vital, often overlooked dimension of the living world.

Fungi are multifarious, showing us that healthy ecosystems take all kinds. They know no borders, insinuating themselves throughout and between overlapping ecologies. Meanwhile, they slip beneath the attention of the wider world even as they link it together, breaking down the detritus of decay and setting the stage for successive generations of life, in often thankless roles that we'd do better to appreciate. In their interstitial existence, they invite us to reexamine how everything connects.

When thinking or writing about such sweeping implications, it occasionally occurs to me to ask: Why fungi? After all, since all life is connected, picking any point of nature to examine should lead one to the same vast and endlessly entangled picture, albeit from a different vantage point. But fungi seem uniquely alluring messengers. They draw us in with their aesthetic subtlety and depth, amaze with their many crucial roles and surprising abilities, all while offering tangible, tantalizing, and often tasty rewards for getting to know them better. On another level, fungi provide potent symbols and examples, resonating with a resurgent impulse to recognize our relationship to—and inseparability from—nature, as dead and dying systems call us all to forge a new way of life.

As I write this, there are open questions about whether the next election for the American presidency will take place. Yesterday, the temperature in Baghdad was 125°F, the hottest ever recorded there.[7] It's no exaggeration to say that we live in a time when our common structures of governance, society, and relations with nature are revealing their failings. Legacy institutions and perspectives alike stand in the way of realizing a sustainable, equitable society. People are hungry for alternatives, and perhaps part of the growing appeal of fungi is that they represent means, examples, and ways of thinking through which we might subvert an unsustainable status quo.

Some of the work and worldviews I document in this book take up fungi as allies in challenging patriarchy, colonialism, capitalism, various extractive or supremacist worldviews that ignore the agency and interconnectedness of nature. Plenty of people in the business of mushrooms are involved because they see a way to make a living, of course, but it is also seen as a line of work that feels connected to virtuous cycles rather than crass accumulation. While there is room for skepticism about what might seem yet another marketing gold rush—mushrooms are also part of a hype-soaked superfood trend, for example, as well as nutritional supplements industry with a concurrent drive to capitalize before research catches up to claims—the communities I've encountered are generally arrayed against these impulses. While some are no doubt motivated by profit to varying degrees, others seem drawn to fungi primarily as a means of democratizing access to food, medicine, even the endeavor of science itself.

This book can be roughly divided into three sections. The first will lay out some of the context from which the new fungi movement is emerging. That means looking at what fungi are, for starters, and how they've been understood (or misunderstood), and regarded (or disregarded) by mainstream science and culture. From there, we'll dive into some of the many facets of the "mycelium underground," from off-grid mycology festivals to mountaintop remediation projects, basement cultivation operations, and online communities that have motivated an international community of amateurs, a term that is quickly and rightly losing its stigma. After that, we look through and beyond fungi to the fuzzier questions of how engaging with unfamiliar, nonhuman agents can inform the ways we engage with the world and one another.

Given that fungi exist everywhere, it's little surprise that this book finds a wide range of people and communities working in a diversity of places: in cities, in mountain forests, in the Amazonian jungle, in the deserts of the American Southwest. If fungi are engaging us in conversation, like a saucer that lands and asks to speak with our leaders, then the delegation that's meeting them is multicultural, multigenerational, multigender, looking to fungi not as a resource to extract or another life-form to conquer but as a powerful silent partner that's been with us all along.

CHAPTER ONE

AMong US

E arth teems with fungi. Throughout forests, jungles, grasslands, and deserts; in puddles, at lakeshores, and on the ocean floor; between cracks in stone and on the peaks of mountains; in all climates and on every continent. Fungi can be found as easily during a walk in rain-soaked woods as in the produce aisle, or simply by jabbing a finger into healthy soil. They are essential and ubiquitous. Turn over a rock, dig under the roots of a tree, scoop up a handful of water, open your mouth: there be the fungi. Stop reading for a moment and take a deep breath—you've just inhaled their spores.

Whether we know it or not, our daily life is rife with fungal encounters: in the beer and wine we drink; the bread, cheese, yogurt, tempeh, and soy sauce we eat; thousands of the medicines and chemicals on which we rely; and the fuzzy splotches that turn our tomatoes to mush.[1] But more than providing conveniences, inconveniences, or culinary experiences, in a meaningful, even literal sense, quietly and largely unseen, fungi bind the living world together. Their exquisitely fine fibers aerate soils, enhancing water retention and bracing against erosion.[2] Meanwhile, fungi churn endlessly underfoot, mobilizing the makings of new life. They are called primary decomposers because they're often first in line to dine on dead or dying trees, leaf litter, and other organic detritus, unlocking nutrients and kicking off the chains of succession that power our planet's ecosystems.[3] Mycological innovator Tradd Cotter uses the term *molecular keys* to describe their ability to unlock a wide range of chemical bonds, such as those that constitute plants, bugs, bacteria, and anything else that lands on a mushroom's menu.[4] In these capacities, fungi connect all living things in essential relational webs; without them, entire ecosystems would collapse.

And yet, while fundamental, fungi are not at the center of things; rather, they exemplify the interconnectedness and interdependence of all life. Our own health relies on dizzyingly diverse communities of microscopic organisms, in what we have come to call our micro- and mycobiomes. Scientists have found that only 43 percent of the cells that make up our corporeal form are actually human; the majority of what counts as "us" comprises bacteria, fungi, and other microbes.[5] For every human gene in our bodies, there are 360 microbial genes.[6] It's enough to inspire an identity crisis.[7] As professor Ruth Ley, director of Microbiome Science at the Max Planck Institute for Developmental Biology, put it, "Your body isn't just you."[8]

Even as microbes have gained prominence in science's view of the world, fungi have remained marginal figures. Fungi were regarded as a funky subset of plants until the latter half of the twentieth century; not until 1969 were they formally recognized as a completely distinct kingdom of life, on par with any other—animals, plants, bacteria—in terms of their scale, variety, and ecological importance. The point is often made that animals, amoeba, and fungi are more closely related to one another than to plants, which may explain something of why they can seem at once strange and uncannily familiar. Many do look like something squarely between animal and vegetable, with an ostensibly rootlike structure underground and mushrooms above that are often described as "fleshy." Some even protect themselves with melanin; leave a shiitake mushroom out in the sun for a while, and its flesh will surge with vitamin D.[9]

The oldest confirmed fungal fossil is dated at about 800 million years old,[10] though it's possible that fungi—and if not fungi, then something quite similar—were found in fossils from 2.4 billion years ago.[11] Regardless, most current views of the evolutionary tree show animals separating from fungi at around a billion years ago.[12] That's around the time when life on earth was still confined to the oceans, and indeed, fungi were at the fore in the move to shore, intimately tied up with the lives of the earliest land plants, in symbiotic relationships that persist to this day.[13] Fossils in Quebec and elsewhere paint the picture of a 400-million-year-old world in which the largest things living on land were the prototaxites, twenty-five-foot-tall spires of what appear to have been a kind of lichen—themselves entanglements of fungi and photosynthesizing algae—that loomed over Ordovician landscapes like blind watchtowers.[14]

Nowadays, plants are the biomass heavyweights of the world, but fungi remain deeply enmeshed with them and their environments, moving nutrients and transmitting chemical information, a sort of circulatory and nervous system in one.[15] As old hands at symbiosis, fungi form networks in a literal sense, as weblike beings below the soil and inside other organisms, and also in a relational sense, serving as interfaces among organisms. All species of plants have been found to harbor what are called endophytic fungi, which live as hidden threads woven in and among their cells—in the roots, stems, leaves, flowers, fruits—serving to metabolize nutrients or dissuade foraging, essentially acting as adopted organs to their host, and vice versa.[16] Meanwhile, the vast majority of plants—some 92 percent of known species—extend their roots' reach thanks to intimate entanglement with mycorrhizae. Literally "root fungi," mycorrhizae solubilize minerals from the soil in exchange for plant sugars produced by photosynthesis.[17]

Yet despite fungi's ubiquity and importance, many people lack even a fundamental understanding of what they are or how they live. As mammals we can't help but have an intuitive sense for what animals are and what's required for our survival: water, food, oxygen, temperatures within certain ranges. Even without any botanical background, many will be familiar with the basics of plants: they soak up water and minerals from the soil through roots, convert sunlight into energy through photosynthesis, "breathe" in carbon dioxide, "exhale" oxygen, and cast cooling shade. These are the barest basics, but it's more than many people know about fungi. Ask someone what a fungus eats and perhaps they'll guess manure, or rotting fruit, or houses, each of which counts as a correct answer. Considering what a vast variety of things fungi consume, though, or can consume, it's difficult to guess wrong; cigarette butts and cicada butts would be equally correct guesses. But ask a stranger *how* fungi eat, and it's a good bet you'll stump them. (Stumps, by the way, are also fixtures of the fungal diet.)

The average person can be forgiven for a lack of fungal literacy. After centuries preoccupied with plants and animals, the institutions of natural science have been slow to prioritize fungi, and few of us receive even a basic education in their biology or ecology. Nevertheless, a great deal is now known, thanks largely to the efforts of passionate mycologists both inside and outside those institutions. Yet many details of fungal biology, their evolutionary history, and their ecological roles in soils, among plants, and in human culture remain cloaked in mystery. For the curious, it offers a lifetime of inquiry and many opportunities to contribute to our understanding of a vital dimension of nature. Luckily for the nonscientists among us, it doesn't require a biology degree to learn about, or from, fungi.

———————

Whether you've tripped on them or over them or just enjoy them on your pizza, most encounters with fungi involve mushrooms. The utterly common supermarket "button" variety may spring to mind, or perhaps the pale little parasols common to fields and suburban lawns. But mushrooms are a universe unto themselves, expressing every imaginable (or, as one soon discovers, often unimaginable) form and texture.

Toadstools peek up from the duff; striated hooves of polypore "shelf" fungi bestud tree trunks; smut and rust fungus appear as dust across vast croplands; pungent, scarlet claws of stinkhorns erupt and unfurl in slow-motion ecstasy before becoming engulfed in flies, or form geometric cages and veils worthy of Buckminster Fuller's fevered dreams. Some look like living icicles or a surrealist's vision of bloody teeth; many appear as tiny cups barnacled upon a twig or blotchy spots on a leaf, never to be noticed by human eyes. Certain varieties bruise in coruscating rainbows when sliced or deliquesce into inky puddles after a few hours aboveground. Many mushrooms might seem soft and squishy, but they can grow with enough force to break through pavement or lift rocks.

Their common names—pheasant's back, witches'-butter, wood ear, elfin saddle—are often so apt that they're impossible to forget once learned. They also comprise a seemingly endless variety of meanings. Depending on whom you ask, mushrooms inspire thoughts of delight, disgust, delicacy, divinity, or death. When encountered in nature, they can seem to vibrate with an inexplicable sense of presence, even selfhood; many mycophiles speak of "meeting" mushrooms in the woods.

But mushrooms are just the tips of a fluffy iceberg, indicating a dense tangle of living fibers below: the mycelium. This is the wispy white patch one sees when lifting a log from putrescent leaves on the wet forest floor or examining the side of a fallen tree. Called the "vegetative" part of the fungus—seemingly a holdover from its inclusion in the plant king-dom—the mycelial mat within which its entire life cycle plays out is often very small, no bigger than an ant or a handful of leaf cells. Others can straddle several acres and live for hundreds, even thousands of years. Upon becoming interested in mushrooms, one often hears about the Humongous Fungus high in Oregon's Blue Mountains; the web of its mycelial "body" extends upward of three square miles, earning it the title of the largest living thing on earth, as well as one of the oldest.[18]

The mycelium amounts to a complex network of cellular strands called hyphae. As the mycelium grows, its hyphae constantly branch and fuse in various directions, searching for food and mates while infusing their environment with a fungal undergirding. Every hyphal tip grows and divides, and those tips do the same again.[19] Viewed up close, and best seen with a microscope or a time-lapse camera, hyphae resemble ice crystals in their constantly expanding search for sustenance, be it from wood, leaves, insects, rocks, bacteria, bone, or symbiosis with another organism.

Along every part of its surface, enzymes ooze out to digest food that is then absorbed back through the walls of the hyphal cells; rather than chew, swallow, and digest their food, fungi flip that script, grow-ing through, dissolving, and absorbing their sustenance. Fungi excrete diverse varieties of exudates, including those that serve to fight off other microbes or attract potential mates as the network of its body expands its reach and negotiates its environment.[20] Since individual hyphae are so thin—down to a few thousandths of a millimeter—the mycelium folds a huge amount of surface area into a small space.[21] As the mycelial web grows from nearly every leading edge, it may add more than half a mile of hyphal length every day. As much as eight miles can be found in a cubic inch of rich, undisturbed soil; under harsher conditions, that length can shrink closer to one hundred centimeters.[22] The incredible fineness of hyphae allows them to wend intricately among and within plant roots, between microscopic cracks in stone, even inside a host's individual cells.

The mycelium is made of millions of hyphal cells, with no distinct center. Split it in half, and the two sections will carry on with their day. Break it into tens of thousands of microscopic pieces, and each fragment will independently resume the search for food and mates, relinking wherever they meet another part of their former, larger self. From tip to hyphal tip, the fungus regulates the flow of energy and information throughout the various zones of its distributed form: Where there's nutrition, energy goes into consuming and distributing it throughout the mycelium; where there are threats, it can marshal an armory of adaptive defenses. And there are always threats. A fungus in nature lives in constant, deadly competition with bacteria, insects, and other fungi that seek to eat them or seize their resources.[23] If pickings seem slim or local conditions take a turn, the mycelium may decide—insofar one is comfortable calling it a "decision"—to retreat, explore new opportunities, or start making mushrooms.

When you look at the forest floor and see mushrooms (lucky you), it's because the mycelium underneath has determined that it's time to cast fate to the wind. Literally, in the form of spores. Mushrooms are also commonly called fruiting bodies, another linguistic relic of their long-standing inclusion among the ranks of plants. Not to be outdone by their botanical namesake, mushrooms take all kinds of curious and colorful shapes that can rival flowers in their beauty and complexity, but their unifying purpose is reproduction, and if we're being honest, that's often pretty obvious just by looking at them.[24]

After the mycelium moves to reproduce, tiny nubs of hyphae called primordia begin to form along the surfaces where it senses light and airflow. Hyphal cells within these putative mushrooms begin to differentiate into the various tissues of the coming reproductive organs, drawing water up from the network as the primordia begin to fill in and unfurl into fully fledged fruiting bodies. Seen through a time-lapse camera, mushrooms often emerge in rolling waves of growth, with a complexity and grace that otherwise eludes the unaided eye.

As they take shape, sometimes over days, sometimes overnight, the fertile surfaces present themselves in a manner that ensures optimal exposure of spore-producing surfaces. And oh do they ever make a lot of spores. Most mushrooms we know best unleash them by the millions, in a process called ballistospory. Like tiny escape pods from a miniature mother ship, spores launch from the mushroom's underbelly in cascading volleys. Peek beneath the caps of mature mushrooms, with the light at the right angle, and you will see them bilge out in vaporous sheets. Cultivators frequently wear respirators while pacing their growing rooms, to protect their lungs from swirling nebulae of spore dust.

Ballistospory is common to the Basidiomycota, a broad category of fungi named for the microscopic organs from which their spores emerge, usually in groups of four, swelling and then bursting free with an acceleration as high as 25,000 Gs.[25] The wild ride is short-lived, since the mushroom cap above acts something like the wing of an airplane, creating a pressure differential. Once the spores encounter the higher-pressure air of the microclimate below the mushroom cap, they instantly decelerate and float off.[26] Spores are so small and light that they can be carried thousands of meters high and travel vast distances.[27]

But ballistospory is just one of many approaches to reproduction. Other mushrooms concentrate their spores internally, either to be released by impact—say, from a falling raindrop—or eaten and deposited elsewhere by a foraging animal, including but not limited to humans. Many animals eat fungi as part of their diet, including at least twenty-two species of primate.[28] Another, much larger category of fungi called Ascomycota is named for the vaguely saclike structures in which its spores are produced. Morels and truffles are familiar ascomycetes, but so are most yeasts, despite being unicellular, generally asexual fungi that neither take the form of a fibrous mycelium nor produce mushrooms. A yeast such as *Saccharomyces cerevisiae*, the trusted friend of beer brewers and bread makers around the world, is classified as an ascomycete. Such taxonomic categories are increasingly porous, though, as genetic sequencing reveals lineages and relationships that run counter to observation and intuition.

There are numerous other types of fungi, with many variations on the themes of sporulation and reproduction. Stinkhorns emerge covered in a smelly, dark paste of spores called gleba, attracting flies that then spirit the sporous sludge off to new territory. The "fire loving" fungus *Pholiota highlandensis* hides its hyphae among the cells of other forest organisms, such as mosses and lichens, before the next blaze destroys its host and offers an opportunity to reproduce.[29] Bread molds and other zygomycetes forgo mushrooms and instead generate spores in little packets right along their hyphal fibers, part of why they're so prolific in our kitchens and bathrooms. Or take *Pilobolus*, which grows atop animal dung in translucent patches of microscopic fur, building fluid pressure in little gelatinous balloons at each tip before launching a spore cap more than two meters away, in an understandable rush to land beyond the "zone of repugnance." Thousands of species called Fungi Imperfecti have lost their ability to reproduce sexually at all. However they do it, though, most fungi we're familiar with have adopted some form of sporulation strategy.

Should a spore meet with the proper conditions for feeding and growing, it begins to germinate. When spores depart from their mother mushroom, they do so with just enough nutrients on board to produce tiny exploratory hyphal threads. Furtively reaching out to assess their new home, the emerging cellular strands secrete pheromones to attract any compatible mates nearby, along with a layer of digestive and protective enzymes. Fungi are essentially inside-out stomachs, and being heterotrophs like us, they have to find and eat their food, unable to generate energy through photosynthesis the way plants do. Many are saprobes, which means they make meals of dead and decaying organic matter, often leaf litter or dead and dying trees. Metabolizing their surroundings and absorbing what they need, they leave behind material that serves as nutrition for organisms further down the trophic chain of succession.[30]

Like plants and animals, fungi are eukaryotes; their cells store and transmit genetic information in the form of densely braided nuclei. That's in contrast to prokaryotes, like bacteria, which carry their DNA in free-floating chromosomes, hence their incredible ability to iterate within their gene pools and evolve at hyper speeds. Inside the hyphae, the nuclei float and flow with other organelles inside a column of pressurized cytoplasm. In more recently evolved species, semipermeable walls called septa divide the hyphal cells like an ever-lengthening train of subway cars. The outer walls of hyphae are incredibly thin, yet can maintain an internal pressure multiple times that of an inflated tire, made possible because of chitin, the tough organic polymer that also makes up the hard shells of most insects.[31]

When a spore forms, it does so in pairs; a single nucleus splits into two half sets of chromosomes, one in each spore. These haploid spores can produce hyphae that will, if they land in a favorable spot, eat and grow and lead much of what makes for a happy fungal existence. But without the other set of chromosomes, it's the end of the genetic line. Should the hyphae of two genetically compatible fungi sense each other, though, they'll reach out to meet and eventually merge, their cell walls dissolving at the point of contact to trade nuclei, initiating a process that cascades throughout the mycelia as two networks become one. At that point, they have formed a new and genetically complete individual (a term worth questioning in this context) that is capable of reproducing. There is no ceremony, not much outward indication at all that something has changed, except in some cases a little clamp that appears between each adjoining hyphal cell, a way of distributing the newly acquired nuclei throughout the network.[32] Otherwise, the fungus just carries on eating, branching, and growing into an ever-wider mycelial web. That is, until it can't, or otherwise feels the time is right to make mushrooms, and the cycle begins anew.

This fungal life cycle and its many variations carry on all around us, all the time; in the soil, under bark, between the cracks in stones, inside our pantries. Fungi can seem at times as much a general force of nature as discrete organisms, working throughout the world as reliably as the pull of gravity. And indeed, it must be so for life to carry on as we know it; they process the dead and dying back into simpler substances for other organisms to eat, mycelial conveyors moving complex organic molecules down the trophic chain, degrading the dead and dying into the stuff of life. Little wonder that in many people's minds, fungi are intimately tied with decay and rot, as if that's a bad thing. Mushrooms are just evidence of that activity, though, and of the fungus living below as it seeks to reproduce. But as its most visible and engaging aspect, the mushroom has throughout human history become a lightning rod for fascination and fear, revelation and revulsion alike.

Philias and Phobias

In the mountains of Central Colorado in the summer of 2019, I joined a caravan of mushroom hunters on a lush slope across from the scree-walled spire of Lizard Head mountain. Someone brought their cattle dog, and as we hoofed up the pine-shaded trail, it instinctively circled us like the hungry herd we were.

We soon dissolved into teams of two. My impromptu foray companion was Ben Lillibridge, an easygoing environmental scientist and mushroom entrepreneur from Hawaii by way of Wisconsin. Together we waded through thick woods and crossed wide-open meadows, stumbling upon an endless wealth of mushrooms that Lillibridge identified for us, utter novice that I was (and am). Phalanxes of scaly brown hawk's wings were especially abundant, and we each chanced upon hefty, squat boletes at the base of some firs, looking like real-life Smurf houses you can eat guilt-free.

Secretly, though, I harbored a personal mission of finding the *Amanita muscaria*, which I knew to be common to the area. Luck was on my side, as a beacon-red flash drew my eye to where thick grass met the brown, needle-carpet edge of the tree line. There, the speckled, forbidden apple-red cap of a newborn *Amanita* was poking through the grass.[33]

With what should have been an embarrassing yelp, I scooped up the mushroom and regarded it with delight before placing it in my bag, no thought whatsoever to what I might actually do with it.[34] By the next day, I'd cooked and eaten the bolete, tossed a *Lactarius* that had gone green, and was amazed to find that the *Amanita* was still warm to the touch, having been actively growing all the while, its cap outstretched and surrounded by a dusting of white spores.

Amanita muscaria, also called the fly agaric, is uniquely charged with spiritual and cultural associations. In the English-speaking world, the mushroom calls to mind Lewis Carroll's apparent reference to it in *Alice in Wonderland*: "One side will make you grow taller, and the other side will make you grow shorter."[35] Its universally familiar shape and color may be the first to pop into one's mind at hearing the word *mushroom*. (As of this writing, an image of the *Amanita muscaria* is what shows up after typing *mushroom* into a Google search, having apparently been designated the Fungi kingdom's official representative.) Anyone who came of age in the 1980s or later may well think of *Super Mario Bros.*, and the power-up that embiggens the pixelated plumber with a satisfying chiptune trill. The emoji symbol for mushrooms is modeled on the *muscaria*. But the cultural associations with this particular mushroom go way beyond video games and internet searches and even literature, offering one example of how deeply entwined fungi are with human culture.

There is almost too much to say about *Amanita muscaria*. Much of the current understanding of its role in history, along with that of most psychoactive mushrooms and substances, amounts to conjecture, but some traditions are fairly well documented. Along the eastern edge of Siberia, local Koryak legend has it that the mushrooms popped up from the drool of Vahiyinin, god of existence.[36] In Croatia, they were believed to come from the bloody spit of Wotan's riding horse. In Central and South America, the fanged Aztec god Xolotl is sometimes depicted with a sagging or dangling eyeball, which some ethnomycologists have suggested symbolizes the fly agaric.[37] If you squint, Xolotl's eyelid does look a bit like the curved cap, the droopy eyeball resembling the distinctive bulbous base shared by family Amanitaceae.

The famous ethnomycologist R. Gordon Wasson (more on him in a bit) was convinced that it represented Soma, a mysterious elixir described in the Rigveda, one of the four sacred canonical texts of Hinduism. He was so confident in having cracked the case that he wrote a whole book on it, with a picture on the cover of the speckled mushroom under the single, red word SOMA. The Vedic texts described the preparation of Soma as involving three filtering steps; Wasson contended that the third and last filter was the human bladder.[38]

Even old Saint Nick carries *Amanita muscaria* associations, vis-à-vis the rituals of the Sámi, indigenous to Sápmi, what we now call Lapland in the north of Finland, not far from where you'd expect to find Kris Kringle's workshop. The account paints a nice picture: A shaman consumes the candy-cane-colored mushroom, perhaps a brew filtered by way of bladder; reportedly, the mushroom was first passed through the bladder of a reindeer, thus strained of its potentially nauseating effects.[39] He enters a snowbound hut through the roof, sharing gifts of visions and riddles and feats of strength or endurance. Similar accounts are attributed to Siberia, but they are also questioned, such as by author Andy Letcher, who suggests our desire to connect the mushroom and the shaman amounts to a kind of Orientalism, a Western idealization of the actual uses and cultural station of mushrooms in unfamiliar (to us) parts of the world.[40]

Many other fungi pop up throughout history and culture. In ancient Greece, for instance, the Eleusinian Mysteries are widely believed to have been facilitated by an ergot fungus.[41] None other than Plato is said to have partaken in these secretive rites, which culminated with the attendants marching en masse to consume a special brew before sharing in a moment of collective euphoria.[42] Demeter, goddess of the harvest, and her daughter Persephone, the goddess whose travels between earth and Hades were said to turn the seasons, were at the center of these rites; a fresco depicting the two handling what appears to be a mushroom is often cited as evidence of its centrality to the ritual.[43] However, ethnomycologists generally credit ergots—not mushrooms, really, but a type of fungus that grows on rye and other wheat, and from which LSD can be derived—as the inebriating agent. The sclerotia formed by *Claviceps purpurea*, a rye ergot, causes trembling and sweating when consumed, along with intense hallucinations, such as the strong feeling of being covered in insects (a phenomenon called formication) a scourge known in the middle-ages as Saint Anthony's fire.[44] Ergotism has also been convincingly linked to the mad behavior reported in the Massachusetts witch trials.[45]

Spiritual and supernatural connotations aside, mushrooms have also offered real utility to humans throughout the millennia. Famous among mycophiles is the discovery of the incredibly well preserved remains of Ötzi (also known as "the Iceman"), who fell dead in the Alps some five thousand years ago. Discovered among his belongings were a pair of polypore mushrooms, the sort that protrude from the sides of trees in hoof-like semicircles, strung upon strands of leather. One is known for its use in reducing gut parasites, and the other as slow-burning tinder for carrying fire.[46] An even earlier example, the Red Lady of El Mirón, was buried more than nineteen thousand years ago in a cave in eastern Cantabria, Spain. Modern analysis of the plaque on her teeth revealed fungal spores, in what is considered the earliest example of mycophagy (a fancy word for "eating mushrooms").[47] Examples are also known throughout North America. The cappuccino-colored, horse-hoof-shaped polypore we now call *Haploporus odorus*, for example, holds deep meaning for the Northern Plains Indians.[48] It's traditionally used as a poultice to stop bleeding, and in scented smoke to treat coughs; jewelry was made from the mushroom, too, as were decorations for war robes.[49] A punk made from another mushroom, *Phellinus igniarius*, has been used for centuries by the Yupik in Alaska as a traditional additive to tobacco, called iqmik.[50]

Mushrooms also hold high station in traditional medicine. In China, for example, *Ganoderma* species—known there as língzhī; in Japan, reishi —have for thousands of years served as a traditional treatment for arthritis, cancer, heart disease, hepatitis, and other ailments.[51] *Ophiocordyceps sinensis*, which grows from the ground out of buried moth larvae, is in some places worth more than gold by weight. Widely regarded as a potent aphrodisiac, in India it is known as keeda jadi, or "insect herb"; in Tibet, yarza gunbu; in the English-speaking world, "Himalayan Viagra."[52]

With mushrooms' myriad uses and effects come just as many possible impressions and meanings. Yet for all that history reveals of humanity's fungal affinity, many do not hold them in very high esteem. Given their uncanny nature as something not quite plant, not quite animal—and now with the widespread connotations of hallucination added to their association with death and decay—it's little surprise that the mere mention of mushrooms can inspire scrunched noses.[53]

When enthralled to mushrooms, a person may naturally assume that they evoke similar feelings for anyone who stumbles upon them. Many people will stop in sublime appreciation at the sight of a passing deer or a beautiful flower; why should it be any different for a mushroom rising up from the grass, a patch of mycelium in the soil, or a slime mold frothing from beneath the bark of a tree? I am, of course, not the first to ask this question. "Birds, flowers, insects, stones delight the observant," wrote nineteenth-century mycologist Charles McIlvaine. "Why not toadstools? A tramp after them is absorbing, study of them interesting, and eating of them health-giving and supremely satisfying."[54] Yet it is a truism among the mycologically inclined that most people reflexively overlook and even revile mushrooms, molds, and others among what are perceived as lower forms of life.

Years ago, on an evening stroll with my college girlfriend, we came upon a patch of common garden mushrooms growing in the grass next to a shopping center. Without warning, my companion revealed her long-standing habit of reflexively kicking mushrooms into oblivion wherever she met them. There wasn't any apparent malice behind the act, but rather total acceptance that this is simply what one does when encountering a mushroom. Certainly, it was impressed upon me and plenty of other suburban kids that any fungus growing in our front lawns was to be avoided, never touched, let alone eaten.[55] But how strange that these humble, harmless, curious expressions of nature could be such magnets for disdain.

Aversion toward fungi seems motivated more by what they represent than by what they actually are or what they actually do. Fungi are associated with rot, decay, and death, which is apt and valid, but they are less often recognized for or associated with their crucial roles in facilitating life. Eating mushrooms carries a sense of risk, or the risqué; molds are seen as sign and symbol of contamination. The more one learns about mushrooms, though, the more unfortunate such examples of myco-phobia become. That's not to deny that moldy food is often best tossed into the compost; but mold is just as often beneficial, hardly deserving of the reflexive disgust it often receives. Such biases present a constant headwind for those working to encourage recognition and protection of these organisms and to elevate their social standing.

———

Few are as familiar with mycophobia as Chilean mycologist Giuliana Furci. Known as her country's first female mycologist, Furci is the founder, CEO, and former chair of a nonprofit called Fundación Fungi, the only international nonprofit dedicated entirely to advocating for fungi at the governmental level.[56]

With a big smile, a curly mane of dark hair, and a natural efferves-cence, a slight accent hints at Furci's London upbringing as she bears passionate witness to the fundamental importance of the Fungi king-dom. (Or rather, *queen*dom, as she is quick to correct. "Language creates reality," she chided at the outset of our conversation, when I mistakenly referred to a fungarium as an herbarium, a common mix-up.)

Furci's message finds a naturally receptive audience among myco-philes, and she regularly speaks at well-attended symposia and events in Chile and internationally. But where she has worked hardest to convey the message of the fungi is in the halls of Chilean government. There, Furci and her colleagues have struggled with—and in many cases suc-ceeded in—the mission of raising awareness about fungi to secure their proper recognition, protection, and respect.

"We like to think we can be a voice for them," Furci told me over the phone as she prepared lunch in Santiago. When Furci speaks about them, fungi are endued with a sense of personhood beyond a mere category of organisms. "It's been a few years now that I have understood that I am some sort of an ambassador. If I walk into a room with politicians I've been working with, they don't see Giuliana, they'll see a mushroom, and that's what I need them to see."

Furci's fondness for fungi began at age nineteen, during a trip to the island of Chiloé, just off the mainland coast of Chile, to study an endangered animal, the *Lycalopex fulvipes*, or Darwin's fox. But the fungal diversity of the island left her enchanted. Spotting a reddish-brown mushroom jutting from a tree trunk, she "wanted to know who it was but couldn't find any literature," she told me, adding that she now believes it was some kind of *Gymnopilus*. "That was the moment. Instant. A lightning ray." Since then, Furci has spent as much time as possible in the forest, or "in their house," as she puts it. "The only explanation that comes close to the feeling is that they choose you," she told me when I asked why fungi, of all things, had become the center of her life's work. "Really, I've tried to do other things, and I just can't. I have an immutable responsibility to the fungal kingdom, or *queen*dom."

The mushrooms seem to know how to choose their allies. In 2010, the foundation fought to get fungi represented in Chile's environmental laws. They succeeded, managing to include language in a law that mandated fungi be classified in a national inventory according to International Union for Conservation of Nature and Natural Resources Red List standards, with requirements for plans to protect, manage, and help in the recovery of threatened species.[57]

In the summer of 2013, the foundation played a key role in triggering the inclusion of fungi in Decree 40, which approved a new environmental impact assessment system that made fungal baseline studies mandatory.[58] Since the decree went into effect, any dam or roadway or other ecologically disruptive development project requires a study of the potential impact on the "funga," alongside flora and fauna—"the three Fs." Chile is the first country in the world to recognize fungi in its environmental assessment standards, a rather shocking fact in light of how important they are.

"That was a result of a really long process," Furci said. "I'm talking years of creating the right scenario for politicians to understand, first, the importance of an *ecosystemic* approach to nature."

Companies that hold sway with the government may have reason to be resistant, after all, if only because accounting for fungi tends to reveal more places in need of protection, which means less landscape from which to extract profit. Passage of these laws also increased the need for establishing baseline studies of fungal distribution throughout Chile, which affords the foundation an opportunity to document its region's fungal diversity. That has led to ongoing field campaigns to educate communities about fungi and train a new generation of Chilean mycologists.

The lack of fungal literacy and representation translates to a lack of space for their advocacy; plants and animals are well accounted for, at the governmental and institutional level as well as in the public mind, with avenues of funding and assessment already established. Advocating for fungi, therefore, is largely a matter of explaining what they are. "That's been the story of our organization," said Furci. "It's like you have ten minutes to talk to someone; you spend nine explaining what fungi are, and then you have one to get to the point. It's tiring."

Fungi are literal grassroots organizers, exemplars of bottom-up processes, and many who work closely with them start to see progress in similar terms. For those mycologists lacking institutional backing and political (let alone social) capital or access to levers of government, a bottom-up mindset seems the natural mode of advocacy for fungi or anything else. But Fundación Fungi takes on the challenge of raising awareness with a distinctly top-down approach. Having originally worked in marine conservation, its founder brings some twenty years of experience in nongovernmental organizations (NGOs) and nonprofits to the task of shifting the perspectives of government officials to "think fungi."

"I know the players very well, on a national and international level, so that made it a lot easier because people tend to believe people more than causes or organizations," Furci told me. "Ultimately what you want to do is translate science to policy. So you need to let the scientists do the science, the politician has to do the politics, and you're an interface in between them. That's the ultimate job of a nonprofit at that level."

Furci's myco-diplomacy was undertaken in the context of a culture that, according to her, largely sees mushrooms as worthy only of derision. Moving governments to take nature seriously can be difficult enough, but in her advocacy for fungi, Furci reported confronting a particularly potent culture of mycophobia. In Chile, she says, mushrooms are often seen as objects of inherent disdain, a tendency that's evident in the language itself.

The Spanish word for "mushroom" is *hongo*, but in Chile people often use the Quechua word *callampa* (k'llampa) instead, a synonym in Chile for "worthlessness." Furci credits the association to the bumper caps you see in pinball machines. "The ball hits those 'mushrooms,' and is bouncing off them and there's loads of sound and light, but zero points," Furci said. "It's 'worth mushroom'; it's worth zero; *vale callampa*." When shantytowns appear, often very quickly and in unexpected places, they're described as *poblaciónes callampa*, or "mushroom towns." The term *callampa* is also common chauvinistic slang for, well, you can probably guess. None of this makes it any easier to get people to take fungi seriously. "Between not being worth anything and being compared with a penis, you have to get fungi in."

The terms used to refer to fungi don't just reflect individual opinion. The cultural regard—or disregard—for such a massive, and massively important, dimension of the ecosystem means that they can't be adequately represented in policy. That's what makes the inclusion of funga at institutional and governmental levels so valuable. If, for example, the World Wildlife Fund were to adopt the three Fs perspective, it opens the possibility of working with them on projects related to fungi, but also *through* the fungi to every other organism that interacts with them. As we've seen, that includes much of life on earth. But the process doesn't seem to work so easily in the other direction.

"If a foundation that gives funding incorporates the right language, you've opened up a whole funding stream, just by changing the words," Furci said. "That's what needs to happen, because today, all funding eligibility for the fungi is made impossible many times over, because the right language isn't in place in the terms of reference; it just says 'plants and animals,' and you're out. In my ten years doing this with the fungi, I've failed to convert marine conservation funders, or plant conservation or animal conservation funders to the fungi. The strategy hasn't worked."

Another common obstacle is sexism. Furci, along with her team, adopted a direct and unapologetic approach to engaging male and too-often chauvinistic interlocutors. "The biggest opposition we faced was the cultural opposition, and that means radically taking on the task, putting the jokes on before they even try to say it," she told me. "If you're a female walking into the Senate, and the senator is telling you he's got a *callampa*, and he's trying to say, 'Oh, I've got a big penis,' you've got to be punk to be able to face that and get change. So we are very—we say *irreverente*—irreverent. We really don't care; we are very radical in our approach and we will take on anybody who has anything to say against fungi in a political arena, whichever arena. We are born from that radical, total, fungal punk.

"The gender issue is huge," Furci continued. "I'm a single mother of a thirteen-year-old boy, and I am a field mycologist. Ultimately I spend most of my time in the forest. And so between the political work and having to be away, you are hard-hit as a female and as a mother in this area of work. A bit like what Jane Goodall had to face with her son."

Indeed, Furci and the celebrated primatologist maintain an open line of communication. Dr. Goodall even wrote a review for Furci's second book.[59] The relationship began when Furci traveled as part of a small group to meet with the famous naturalist, and took the opportunity to hand over a copy of her first book.

"We saw each other again the next day, and she came up to me and she said, 'You're with mushrooms where I was with the chimps when *I* began. Don't stop, don't stop.' She was very, very forceful in telling me not to stop." Furci heeded her advice.

In September of 2020, the foundation expanded its remit from Chile to the globe, achieving nonprofit status in the United States and gathering a board that includes the world-famous mycologist Paul Stamets and actor Nathalie Kelley. The inclusion of media figures among mycologists and scientists reflects a renewed effort to influence the language used with respect to fungi. But the goals are bigger even than that now.

"Our work isn't always focused only on governments, but we will be engaged with political endeavors that for the first couple of years have to do with language, have to do with the three Fs," said Furci. "Eventually, we want to be able to move to support new initiatives, all the support we've never had! It's the first global NGO for the fungi, which I think is more than deserved for them."

CHAPTER TWO

Scratching the Surface

F ungi don't live their lives in a vacuum, even if some can survive in space.[1] Quite the opposite is true. They are inextricably engaged with the world around them, both as a matter of individual survival and as performers of fundamental functions within their host ecologies, often in symbiotic relationships that can muddle the question of who is host to whom. Lichens are a prime example, each an intimate intertwining of fungus and algae. The lichen's physical shape is defined by its fungal aspect, while the color and photosynthetic abilities are provided by chlorophyll-containing algae or cyanobacteria enmeshed throughout. The very word *symbiosis* was coined to describe lichens.[2]

Fungi also perform a vast range of roles throughout the world's soil ecosystems. They're "the egg in the cake," as mycologist Giuliana Furci puts it. Maybe the most compelling example of this integrative function can be found in the rhizosphere, the hyperactive zone of life that swirls around the roots of plants. Soils are the most biologically diverse places on earth, and within them, the realm of the rhizosphere is among the most complex and dynamic, defined by a microscopic froth of biological, chemical, and physical interactions among plants, animals, and microbes. It's a dizzying commotion of life that's bound up in the hyphae of mycorrhizal fungi.

At the University of California, Riverside, PhD student Danielle Stevenson greeted me under the towering carillon at the center of campus. A short walk took us to the lab where she was studying arbuscular mycorrhizal fungi, or AMF.[3] Found in environments ranging from grasslands to deserts to temperate and tropical forests, AMF partner with upward of 80 percent of all plant species, and account for between 20 percent and 30 percent of the microbial biomass in soil.[4] AMF insinuate themselves within the very cells of plant roots, most often in mutually beneficial relationships that form the basis of a kind of underground energy economy.

Ectomycorrhizal fungi, or EMF, are the other main group, found mostly in boreal, evergreen, and temperate deciduous forests. They produce profuse filaments that reach deep into the soil to help the host plant take up water. Instead of breaking into the plant cells, they create a mantle that envelops the root in a kind of fungal-fiber stocking. Beneath the surface of the interwoven sheath, hyphae surround and penetrate the root's first few layers of cells, spreading out to form internal layers that regulate the plant's hormones and guide the behavior of the plant. EMF can, for example, influence trees to grow fewer root hairs. They also produce most of the tree-associating mushrooms we eat, like porcini and chanterelles; AMF, by comparison, don't produce any mushrooms.

From the surrounding soils, an AMF mines phosphorus, nitrogen, and other trace minerals, which it trades for upward of 20 percent of the sugars and carbohydrate energy its host plant generates through photosynthesis.[5] Fungal hyphae are generally much, much thinner than plant roots, and so the former expand the latter's surface area by anywhere from ten to one hundred times.[6] Trees with mycorrhizal associates have been found to contain nitrogen and phosphorus in amounts as much as double those without. Many fungi that connect with plants this way, and those that later eat the plants as they decay, also sink significant amounts of carbon in the soil.[7] Complicating the picture, some fungi can actually switch roles from mutualistic to parasitic and saprobic as a host declines.[8] Most significant to Stevenson's work, though, were the ways these fungi interact with potentially toxic elements and heavy metals.

UC Riverside has its, well, roots in agriculture, beginning as the University of California Citrus Experiment Station in 1907. Stevenson's work there was conducted within the lab of the Soil Biogeochemistry

Group, part of the department of Environmental Toxicology, largely focused on contaminants in farmland or groundwater. "They call it the dirty people lab," she joked as she fired up a microscope on the lab bench, overlooking the sunny courtyard through crowns of eucalyptus. "I'm the weird one who does microbial stuff." Her research involved sourcing metal-contaminated soils from sites all over the country, using genetic sequencing techniques to determine what, if any, AMF were present, and whether they might mitigate toxic metals from contaminating their host plants, with an emphasis on food crops.

Sun-seared from time spent outside and endlessly excitable about her subjects of study, Stevenson had moved to California from British Columbia just a year prior, where she was involved in community organizing around regional food security and regenerative soil practices. In so doing, she struck upon a deep passion for fungi, which turned into a small business and ultimately the subject of her doctoral studies. Despite the narrow focus of her academic work, Stevenson's curiosity for fungi was expansive and always active.

"I had been working with plants and composts and soils and stuff, but I didn't know anything about fungi," she said as we reviewed potential root samples to place under the microscope. Since then, it was hard not to see fungi everywhere. One of the first things she did upon arriving in LA was to test for them in samples from the La Brea Tar Pits—sure enough, they were there. It was a widely popular speech given by famous myco-evangelist Paul Stamets, called "Six Ways Mushrooms Can Save the World," that first drew her attention to fungi, along with countless other recently minted mycophiles, including myself.[9]

"I saw that TED talk—you know the one, of course!—and I was, like, 'Wait, whoa. Maybe I need to learn about fungi, because they can maybe help decontaminate soil.' And then I was, like, 'Oh, wait. To be able to do that, you have to know how fungi live, and how to grow them and take care of them.' So I started doing that, and then [my company] DIY Fungi happened, and I had a spawn business accidentally, and then I was teaching people about all this stuff."

Stevenson guided me through the process of setting up glass slides, and it quickly became evident that I hadn't practiced the skill since high school. Clumsily, I placed onto each slide a small wet sprig of root sample plucked from a jar of preservative fluid. "Where did I get this tomato?" she wondered aloud about our first subject. "Oh, from my own garden!" Dialing in the viewfinder, she sought out the signs of mycorrhizae hidden within the knotted strands. Visual identification of fungal filaments is a traditional microscopal skill that remains valuable in the age of computer-mediated genetic analysis that has come to dominate many aspects of the natural sciences. Stevenson spoke reverently of her recently retired mentors and "mycorrhizal grandparents" at UCR, Edith and Michael Allen, who could identify specific AMF species just by looking at their hyphal threads through the microscope. "I'm not at that level yet," she noted deferentially.

Each sample had been bathing in a special dye designed to stain fungal cells, "like a highlighter that shows you in blue the fungal cell components." The result was that the magnified roots appeared nearly translucent, interwoven with the fibrous, cerulean tendrils of AMF hyphae. Stark against the pale white backlight, some hyphal threads followed the contours of the root and its cells; others pointed straight out at perpendicular angles. These were functional extensions of the roots, knitted among them and along their length to form a living fabric of reciprocity. After a few minutes of microscopic exploration, we traded out the tomato plant for roots from a palm tree, and quickly found a similar image. When we zoomed in closer, the insides of the plant cells came into focus, wherein the fungi had formed tiny tree-shaped knots called arbuscules, acting as the interface between fungus and plant. Pausing to behold the image of a microscopic tree within a tree, I unconsciously whispered a drawn-out *whoa*, which caused Stevenson to chuckle. "That's what I sound like at the microscope," she said.

Mycorrhizae are fundamental to ecosystems throughout the planet. Moving carbon and nitrogen between plants and aiding in mineral uptake, the associations they form are often highly mutualistic and sometimes even obligate, a term that describes relationships in which one partner can't survive without the other.[10] Orchid seeds, for example, can't germinate without a fungal partner.[11] Codependency isn't just a human trait, it seems. "Sometimes in highly managed farms, or landscapes where there's a lot of fertilizer applied all the time, there aren't as many mycorrhizae," Stevenson explained. "The plant is less incentivized to form a long-term relationship where it's going to have to give something, when it can get what it needs for free somewhere else."

Numerous mycorrhizae can associate with a single tree, or with several, connecting plants in what nerds often refer to as "the wood-wide-web." These dynamic underground linkages can be quite large; some species of fungi coil their hyphae into long-distance rhizomorphs that can stretch as far as twelve feet from the host tree, linking to the roots of various neighboring plants.[12] Older trees are usually hubs for the most and greatest variety of fungal connections, dubbed "mother trees" by forester and biologist Suzanne Simard.[13] One tree can harbor as many as two hundred different fungal species, suggestive of just how complex the relational possibilities are.[14]

The role of mycorrhizal fungi in helping plants thrive is complemented by their ability to hold soils together, and for regulating the movement of toxic heavy metals into and around their plant partners. "With AMF, it's super cool," she said. "They secrete huge amounts of glomalin, which is a protein that does all this cool stuff, but it also complexes with metals, and binds them outside of the fungal cell wall. The metals basically get stuck so that they're not going to move around and cause problems." Glomalin serves other purposes, too. For one, its presence is a sign of healthy soil, in which it works as an adhesive agent of sorts, particularly in earth that remains undisturbed by tilling. The sticky glycoprotein helps retain moisture, resist heat, and stores as much as 15 to 20 percent of the soil's organic carbon.[15]

AMF also have a variety of tricks up their sleeves for dealing with threats to their or their host's well-being. Some will even take on metals mixed within the soil as part of their own defenses. "A lot of fungi living in soil are prone to being attacked or eaten in some way by other microbes, or worms or nematodes, and copper is like an insecticide," said Stevenson, clearly enamored with the sheer cleverness of these fungi. "They actually glom it on, they 'paint' themselves with the copper as a form of shielding, making themselves toxic to possible predators."

Sometimes, mushrooms produced by EMF also have the ability to absorb and store metals in their flesh, preventing their absorption by the associated plant, or into water flows. That means when you spot a bolete or a porcini or some other tree-associating mushroom, if you have any plans to eat it, you'd better make sure it's not downslope from heavy metal contamination. Certain EMF also exude enzymatic cocktails that create a barrier protecting the root from bacteria and competing fungi. For example, should a pine tree's nitrogen start running low, its associated *Laccaria bicolor* fungus will release a toxin that kills any springtails unlucky enough to be nearby, conscripting them into the role of emergency fertilizer.[16]

The remaining mycorrhizal families exhibit variations on these themes. The mycorrhizae that germinate orchids constitute their own type. Monotropoid fungi form associations with nonphotosynthesizing plants, like ghost pipe; the fungus is hoodwinked into providing nutrients to the plant, which can't provide anything in return. Often, the monotrope draws energy gathered by the fungus from some other, more reciprocal plant partner.[17]

Stevenson's excitement about fungi is understandable. Serving so many crucial and often surprising ecological roles, mycorrhizae in particular are powerful examples of the reciprocal, interdependent, diverse communities that make up a healthy ecosystem. It might seem counterintuitive to look to fungi in hopes of lessening contamination of croplands, but not to Stevenson. "As a synthesist, and someone who sees everything as interconnected, it totally makes sense to be studying toxicology at an agricultural school, looking at AMF in a soil biogeochemistry lab," she said at the end of our visit. "I think it's recognized that science is becoming more interdisciplinary, and there's a benefit and value to that, but it's not fully there yet."

Indeed, fungi have traditionally been studied and assessed quite narrowly as forest pathogens. For all the necessary roles they play, after all, fungi are competitive and opportunistic just like every other living thing. If their opportunity comes at the cost of something of value to someone else, they prove formidable foes. Fungal blights can take down hectares of forest just as surely as a house. Crops in particular must be protected from fungal infections, as the Irish Potato Famine of 1845 demonstrated in spades.[18] Yet even pathogenic behavior has value within mature and healthy ecosystems.

Meet the Beetles

The dying lodgepole stood on the edge of a narrow clearing just uphill from Teapot Lake, one of several tarns smattered among the Uinta Mountains. Taken in by the sinuous horizon of mountaintops that surrounded us, I wouldn't have noticed the conifer if my hiking companion, Rob Filmore, hadn't pointed it out. A talkative retiree and local field guide who hiked the area several times a week, Filmore was taking the opportunity to note the trees we'd be passing on our high-altitude mushroom hunt. Following contours and seeking out pines and fir, our hope was to hit upon a patch of porcini, mycorrhizal fungi known to associate with those trees. Rob also hoped to impress upon me the scale of the impact made by bark beetles, which had ravaged Utah's forests in recent years, along with vast swaths of the North American West.[19]

The lodgepole looked like the center of some sort of woodland crime scene. Its surface was riddled with pitch tubes, entrance wounds encrusted with the sap and resin that had oozed out. The tree was mostly denuded of needles, its bark dry and pallid, the ground around it dusted with frass left behind by the burrowing bugs. My involuntary reaction was to groan in sympathy, but Rob was quick to reassure. "It's natural," he said. "Once these trees go down, others will have a chance to come up." Indeed, dozens of saplings basked in a column of sunlight opened around the declining pine. But around us were other trees in a similarly sad state, and turning to look at the quilted patchwork of withered stands on a hill across the meadow, I found it hard not to feel alarmed at the sheer scale of ecological evisceration wrought by these minuscule bugs.

Roughly the size and appearance of mouse droppings, mountain pine beetles have become a scourge throughout North America, from the Yukon down through the Rockies and even into Mexico. According to the US Forestry Service, between 2000 and 2018, *Dendroctonus ponderosae* ransacked stands across 26.7 million acres.[20] The footprint covered by all varieties of western pine beetles accounted for fifty-nine million acres, an area roughly the size of Utah. Concerns about their impact extend well beyond trees. In Yellowstone National Park, just a few hundred miles north of the Uintas, enough whitebark pine was affected by mountain pine beetles to raise alarms about the prospects for grizzlies, nutcrackers, and some twenty species of birds and small mammals under threat of losing their habitat.[21]

As with every other dimension of the living world, climate change will certainly prove a factor in the frequency and effect of bark beetle outbreaks. Dried out trees have diminished ability to defend from attacks, and with increasingly frequent droughts, more trees become vulnerable. The beetles thrive in warmer conditions, too, and so emerge each summer in greater numbers, compounding feedback loops that don't break until it gets too cold, or when the invading insects have nowhere left to eat and breed. That is, when the trees run out. Little wonder the mention of bark beetles inspires visions of roiling clouds of gnashing mandibles turning green hillsides into arboreal boneyards. But many people are just as concerned about the bugs' voracious, toothless passengers. "You know what kills the tree," Rob remarked as we walked on. "It isn't the beetle, it's the fungus."

There's no doubt that the health of certain fungi means the death of or discomfort to something else. Remember the Humongous Fungus from chapter 1? It's a species of *Armillaria*, or honey mushroom, considered a forest pathogen. The fungus forms thick ropes of rhizomorphic hyphae that gradually choke the coniferous and broad-leaf plants they infect, exhibiting a dark, tough texture that's earned it the nickname "shoestring root rot."[22] The splotches of sickened stands infected by this fungus in Oregon's Malheur National Forest are so large their size is best assessed from the air. It's just one of numerous examples of why fungi have long been seen by foresters first and foremost as parasites. Although bark beetles aren't known to carry *Armillaria*, they have evolved associations with numerous other tree-intruding fungi, relationships that demonstrate the endlessly integrative roles fungi play in their ecosystems.

Sometimes the associations are quite loose; a beetle's exoskeleton is prone to become dusted with the spores from any of a number of different fungal species. But many beetle-fungus relationships are highly sophisticated and specific. *Ophiostoma novo-ulmi* and *Ophiostoma ulmi*, for instance, the fungi behind Dutch elm disease, are closely associated with the banded elm bark and European elm bark beetles.[23] In some of the most refined cases, beetles transport the fungus from tree to tree in specialized pockets called mycangia, which can be so fine-tuned that they include special glands that nourish the fungal stowaways until they can get a foothold inside the tree. With these adaptations, the result of millennia of coevolution, it's fair to say the fungus and beetle are made for each other.[24]

As a pioneering female burrows in, she dispenses a pheromone that invites a fusillade of fellow beetles to perforate the bark in search of a cozy place to lay eggs.[25] They find it by chomping their way into the phloem, the tree's nutrient-conducting tissues, wherein sugars circulate from the photosynthesizing needles. It's in this sensitive cambial sweet spot that the beetles carve out their galleries, woody catacombs in which their larvae develop and feed. The tree protests the incursion by releasing flows of defensive resin as the bugs burrow in, an attempt to eject, immobilize, or kill the invaders. If enough beetles keep up their incursion until the tree's defenses are exhausted, it's game: bugs.

In the wake of all this infra-tree commotion, the hitchhiking spores germinate, their hyphae creeping into the wood. In contrast to mutualists like mycorrhizae, these fungi could be considered parasitic or pathogenic, as they make a meal out of their host. It's commonly held that a beetle-borne fungus fulfills its function when it kills the tree, mainly by disrupting circulation. So common is this perspective that it's been dubbed by some within the field as "the classic paradigm."[26] The view is consistent with a widespread view of fungi as forest pathogens, but according to Dr. Diana Six, researcher of bark beetles and fungal interactions, and chair of the Department of Ecosystem and Conservation Sciences at the University of Montana's College of Forestry and Conservation, the classic paradigm doesn't tell the complete story. "For the most part, these fungi of the tree-killing beetles are not strong pathogens," she told me. "They can't really kill trees."

A series of papers, including those written by Six and her colleague, tree pathologist Michael J. Wingfield, have argued that the classic paradigm is in need of an update, pointing to what they see as fundamental flaws in the theory. For example, in trees killed by the beetle *D. ponderosae*, considered a premier pest in the forests of the western United States and Canada, the presence of tree-killing fungi is rarely detected. In fact, according to Six, the most virulent fungal passengers of the most aggressive bark beetles are rare or missing in many populations of beetle-killed trees.[27] Another area where the classic paradigm seems iffy is with regard to the time between the beetles' arrival and a tree's decline in health. The moment of death can't be pinpointed in trees quite the way it can for animals. Instead, trees cross a point of no return, a threshold that according to Six occurs well before the fungi are even up and running. "These beetles get in and they have to overwhelm the defenses of the trees within a few days, and if they don't, the tree kills them," Six said. "It takes a couple of weeks before they actually start growing, and by that time, the tree is already declining."

Rather than killing the tree and weakening its defenses, Six suggests, the fungi are using their talents for chemistry and moving nutrients to feed beetles. If some fungi benefit the plants with which they partner, it stands to reason that they can do the same for their beetle associates. The latter's larvae are usually larger and more likely to complete their development when fungi are involved; they grow more quickly, too, and have a higher chance of reproducing than bark beetles without fungal fellow travelers.[28] The closest associations are obligate; the beetles can't live inside the tree without their fungi. Meanwhile, the fungi benefit in the form of transportation, persisting generation to generation by hitchhiking from one new trunk to another.

"When fungi get into the tree, they grow deep into the sapwood, gather up nitrogen or phosphorus and then pump it out to where the beetle larvae are feeding, and that allows the beetles to essentially live on wood, which is a really crappy food source," explained Six, adding that, upon testing this notion, the evidence to support it became clear. Nitrogen content arriving to the beetles went up by 50 percent in the presence of the fungi, and phosphorus by several hundred percent. "These things that were otherwise almost nonexistent in what the beetles were feeding on, we were finding it being moved from the inside of the tree out to where the beetles were feeding, and it was the fungi that were moving it. Fungi are really good at moving stuff."

So if tree death is not the result of a fungus clogging up the works or beetles sapping defenses, what actually kills the tree? According to Six, that's the 64-million-dollar question. "Nobody's figured out a way to actually physically do the test," she said. "You can't just go up and punch thousands of holes in the tree and then see if that does it, because there's actually tunneling that occurs after the attacks, and you can't mimic that without putting beetles in. And then *they've* got fungi, so you can never get them fully sterile. You kind of have to go by circumstantial evidence, and that is that the trees' physiology basically quits defending; they quit making the defenses, they quit pushing pitch out, there's a big drop in resin flow. And this is in a few days, so the fungi have barely even been put into the tree yet. They're not growing and not clogging up its systems, that's for sure."

The picture of bark beetle ecology appears at first blush to be one of linear destruction: more bugs, more problems, or at least fewer trees. And among the many players in bark beetle ecology, the trees always seem to get the short end of the stick, as it were. But there is a dynamic of feedback and adaptation among the beetles, their symbionts, and even their victims. Indeed, much in the same way that fungi might not be the agents of arboreal death that they're often assumed to be, the bark beetle may not quite mark the imminent demise of the forests in which they swarm.[29]

Through her research Six has come to hypothesize that, over the long term, bark beetles' activities may actually be helping forests adapt as the climate changes. "In places that got hit really hard quite often [by beetles], there were still like five, seven percent of the mature big trees still standing just fine, virtually untouched," she said. "We started wondering, why is it that these guys are fine, when there were literally billions of beetles flying around with nowhere to go because they'd eaten themselves out of house and home, and yet they didn't touch these trees? So we did some genetic screens, and the trees that survived are genetically different from trees that are susceptible to the beetles."

Some inherited trait or traits among the surviving trees seemed to be making them more resistant to bark beetles and the conditions in which they thrive, which tend to be warmer and drier. Therefore, the theory goes, culling trees that are not genetically suited to the shifting conditions leaves behind trees that are, setting the stage for better adapted forests in the future.[30] That is, assuming those trees aren't wiped out by, say, loggers. None of this is to say that forest managers ought to let bark beetles run rampant, of course, but it might suggest a reconsideration of how their ecology is assessed in relationship to forest health, and the role of the fungi in that picture not as killers but mutualists. It also shows the value of outside perspectives even in well-established fields.

Six is an unabashed fan of fungi. She's even brewed her own beer from fungus cultured from a beetle.[31] Following a challenging childhood and adolescence, she became the first in her family to go to college, studying microbiology and integrated pest management, earning a degree in veterinary entomology and ultimately a PhD in entomology and mycology, followed by a postdoc in chemical ecology. She credits her clement conclusions about fungi to a mycological education that took place largely outside the common framing of forest pathology—if you're a microbe, one surefire way to become the subject of intense study is to pose an economic threat to timber board feet.

But Six didn't approach fungi as pests, so she didn't analyze them as such. Instead, she read outside the forest pathology literature, including about the mutualism between insects and fungi. Leaf-cutter ants for example, are well known for actively cultivating and harvesting fungi in little gardens, dutifully foraging and composting chunks of leaves to feed the mycelia. In exchange, the fungus helps fend off other microbes while also providing food for their colony.[32] Termites farm fungi, too, tending to them in fuzzy "combs," where they thrive by digesting the wood and rendering it more edible for the termites. Eventually, the *Termitomyces* fungus sprouts giant mushrooms from the termite mound, which people above happily pick and eat.

Incorporating these perspectives led Six to a sense of the symbiotic, generative dynamics at play between beetle and fungus, and within the ecosystem as a whole. Instead of seeing them simply as twin prongs in a deadly attack on trees and forests at large, she saw them as partners keeping each other alive from generation to generation. "It just seemed kind of obvious that that's what was happening here," Six told me. "If we hadn't gotten away from the idea that these things kill trees and looked at something else, like maybe the beetles are using them for food, we could never have linked them into energy flows in a food web, in an ecosystem." Things are never quite so simple as saying that one organism kills another anyway, because at the level of ecosystems, death ultimately serves life.

CHAPTER THREE

A Neglected Megascience

I was in the pristine Uintas as the guest of a group of undergraduate biology students from the University of Utah, part of a mycology class that had come to the mountains to find and identify mushrooms in their natural habitat. Rounding the southern corner of Teapot Lake, I soon found them beside a wooden fence, its beams engraved with the meandering galleries of bark beetles, excitedly passing around the day's finds. Their professor, Bryn Dentinger, listened intently as one student described the features of an interesting lichen she'd spotted. "Did you grab that?" he asked. A hesitant no was the answer. "Bummer," he said. "They're fun to look at microscopically."

The group's buzz of energy may have been due to the simple joy of being outside for a few hours, or spotting a bald eagle flying to its nearby nest. Maybe it was related to a sugar rush from the doughnuts Dentinger had bought for everyone en route to the woods, a field trip tradition. The pastry boxes were quickly emptied, subsequently used by students to tote their mushrooms, using the squares of wax paper in place of parchment for wrapping them. Several had encountered remarkably rich pickings around Teapot Lake, returning with colorful varieties that their teacher encouraged them to sniff as part of the identification process.

Dentinger, an accomplished mycologist in his early forties, first got interested in fungi as a child. "My mom gave me a mushroom guide and said, 'Hey, can you tell me what these mushrooms are in our yard?'" he recalled over the phone before our hike. "I spent that afternoon looking at them, and observing their subtle features for the first time, and seeing their intricacy in a way that I had never observed before, and I got completely hooked by it. So this whole world was just unveiled to me, and it's fascinated me ever since. And I think the thing that really reinforced that fascination early on is that, when you go on a foray, it's a treasure hunt, and what kid doesn't like to go on a treasure hunt?"

In terms of treasure, the haul from the Uintas represented a bounty, and a sight for sore eyes. Conditions in the area had been extremely dry the two years prior, meaning scant pickings for mushroom hunters. At certain points, things became so dire that Dentinger called off class collection trips entirely. The year before, he paid a colleague in upstate New York to go into the field and collect specimens, then FedEx them to Utah so the students had fresh material to study. The week before my arrival had brought significant rains, though, and the boxes brimming with colorful mushrooms marked one of the most productive class outings yet. Encouraged by their success around Teapot Lake, Dentinger drove the group to a favored spot of secluded woods above an unpaved access road. There, another estimable haul of fingerlike coral fungi; cartoonish russulas with candy-red tops and milky-white underbellies; a young silver amanita, its clenched cap yet to unfurl into the shape of a wind-inverted umbrella. I plucked a monotrope from the ground, one of those nonphotosynthesizing plants that "trick" mycorrhizal fungi into sustaining them, and considered it an act of helping some unseen fungus out of a one-sided relationship.

A quiet hour's drive in a minivan freighted with fungi and napping undergrads took us back to the University of Utah. There, the students rallied in the lab and set about documenting their collections according to a set of handout forms. There were many details to consider. Did the mushroom's cap have a conical, planar, hemispheric, or convex shape? Was its top smooth? Scaly? Shaggy? Were there neat rows of gills underneath, or a spongelike surface of pores? Did the edges of the cap wave, undulate, or crack? How large was the stipe (the mushroom equivalent of a stem); where and how did it connect with the cap; what kind of texture did it have?

Experienced pickers take thousands of such features into account when identifying a mushroom, sometimes intuitively, sometimes using detailed keys. In addition to outwardly visible traits, they will also judge the color of spores (easily collected in piles by leaving the cap of a mushroom gills-down on a piece of paper for a few hours); the season; weather conditions; the type of wood, debris, or other substrate from which they grow; whether the mushroom flesh bleeds or bruises when cut, and, of course, the location where they were found.

Features visible only under a microscope are also essential to identification, as are the various and often beautiful microscopic details of spores themselves. Smell is a major factor in mushroom identification, too, although words tend to fail when trying to describe their aromas. The nose is highly suggestive, after all, so it's one of the more subjective dimensions of mushroom identification. For many mycophiles, sniffing a mushroom is as much about feeling a closer connection to a specimen as it is a matter of identification anyway.

While students cataloged details, Dentinger began assessing which specimens were worth saving. Those deemed uninteresting were tossed into the trash while the rest were cut into halves and stacked on drying racks. In most cases a mushroom passed muster simply because it wasn't already represented in the university collection, kept at the museum just uphill from the campus. "I am accepting everything that is not decomposed or moldy at this point," Dentinger told me. "We are starting from scratch, so it's not like we have any limitations on space just yet, and at this scale that's not really an issue."

There is certainly no risk of cataloging the entirety of the fungal diversity in Utah, or anywhere else, for that matter. Fungi are known to be incredibly diverse, but to what extent remains an unresolved question. Current estimates place the total at around 3.8 million species. Yet only some 144,000 have been identified, less than 10 percent.[1] Prior to genetic sequencing techniques that made it possible to identify a wide range of organisms in organic complexes such as soils, the fungi most often identified were those that produced mushrooms, because, well, you could see them. But many fungi don't produce mushrooms at all, or live invisibly and in unexpected places. The only way scientists ever find them is by grinding up a sample of, say, soil or dung or plant matter, and then looking for the telltale genetic signs of any fungi in the mixture of organic material. For mycologists, these "cryptic" fungi represent a vast, enticing unknown dimension of the living world.

Genetic sequencing has also exploded the understanding of fungal taxonomy. Mycologists and mycophiles are finding themselves relearning long-standing scientific names for mushrooms, which are constantly updated as genetics reveal new truths about their relationships. Many mushrooms share similar physical features, and are therefore naturally lumped together by taxonomists, only to be revealed by their DNA as actually quite distantly related. This outcome is often chalked up to convergent evolution, instances when two organisms look to be closely related, but actually aren't. Instead, they've arrived at similar outcomes by way of distinct evolutionary paths. Hence, there are "false" varieties of morels and chanterelles, for example, which at first glance look just like their delectable namesakes but can be distasteful and even toxic.[2] This has thrown the traditional taxonomic relationships, based on morphological (that is, observable) details, into a kind of chaos.[3]

It's not just their morphologies that converge but also their biochemistry. Mushrooms that produce psilocybin and psilocin, for example, do so via at least two evolutionary pathways.[4] It works the other way, too. *Divergent* evolution leads two closely related mushrooms to look very different. For all the revision this has caused within the standard taxonomical naming system of Latin binomials, the trade-off is a much more complete if also far more complicated picture of the evolutionary relationships and lineages. Much has been learned about fungi by wandering the woods and looking under a microscope, but searching for insights in the genetic realm now takes up a growing share of the formal science of mycology, and it is constantly reshaping the scientific picture of how organisms really relate and function in ecosystems. "God created; Linnaeus organized," the latter was known to have said of himself. A grand claim increasingly undermined by another, invisible, three-letter agent credited with creation: DNA.[5]

———

The Natural History Museum of Utah (NHMU) is an earthy, angular building that appears as much the result of tectonic activity as a work of architecture. With a commanding view of Salt Lake City and the strip mine–scarred Oquirrh Mountains across the valley, it sits right along the former shoreline of the ancient Lake Bonneville, on a ridge where narrow creek-carved canyons slither down from Lookout Peak. Bryn Dentinger met me at the museum entrance as the sun was rising, leading us through the central atrium, a dark and echoing space evocative of caves and cathedrals. Dinosaur skulls hung beside a tall glass wall displaying Indigenous art and artifacts. Above us on the second level, a rocky bridge led to an exhibit on Great Salt Lake, while the pictorial forms of plants crept along the walls in projected light.

A security elevator took us up to the lab where Dentinger and his graduate students work on fungal phylogenetics and metagenomics. That is, teasing out genetic information from the myriad organisms mixed up in soils, dung, and other complex samples, then mapping out their relationships to one another. In Dentinger's case, that life was fungal. By documenting and comparing the genetic "bar codes" ascribed to different fungal specimens, his team could gradually reveal their evolutionary relationships, in particular those of mycorrhizal varieties growing with trees in the Guineo-Congolese rain forest. One question they hoped to answer was whether African mushrooms shared common ancestors with South American mushrooms, since the two continents were once conjoined hundreds of millions of years ago.[6]

This data-intensive deep dive into evolutionary history was carried out in a space about the size of a small studio apartment. Laid out in the shape of a horseshoe, the lab was stuffed with thermocyclers, computers, centrifuges, and other high-tech hardware. Along the desk, withered clusters of mushrooms lay on drying racks, fruits of a recent collection trip in Cameroon. The lab was located on the same floor where all the museum's dried biological collections were kept, which meant no food, water, or biological material of any kind was allowed up without first being frozen for two weeks, measures taken to prevent any hitchhiking arthropod eggs from threatening the hundreds of thousands of specimens stored just around the corner. "We don't have an option to process our material anywhere else, so we currently just bring it up here," Dentinger admitted with a chuckle. "Nobody's said anything to us yet."

For all the sophisticated technology and analytical techniques involved in teasing a fungal tree of life out of DNA molecules, the process starts with the comparatively simple act of picking a mushroom. Drying them out preserves identifying details, including at the microscopic level. For later study, the samples can be rehydrated with a solution that essentially inflates their cells. Until then, in theory, these mushrooms would stay preserved forever. "I think we're well placed in Utah to be custodians of this material," Dentinger said, "because we don't have to work hard to keep things dry."

Dentinger had achieved as much in the field of mycology as anyone could hope to at his age. After earning his PhD in plant biology, he researched methods of DNA bar coding at the Royal Ontario Museum in Toronto, moving to Oregon to study the evolution of *Dracula* orchids, which, amazingly, mimic the look and smell of mushrooms. He then accepted a job in London as the head of Mycology at the prestigious Royal Botanical Gardens, Kew. Two years prior to my visit, he and his family had moved to Utah for his job at NHMU, where he occupied a dual role, both as the curator of mycology at the museum, and as associate professor at the University of Utah, just down the hill. The new gig had meant adjusting to a new environment, professionally as well as geographically.

When Dentinger first interviewed for the position in 2016, a curator disclosed never having actually heard of mycology before; since then, some members of staff repeatedly used the term *malacology* to refer to his area of expertise. It's a sign perhaps that mycology remains something of a misfit among institutions of the natural sciences. "We have to constantly explain the importance of this to our colleagues," he said.

Dentinger's pursuit of his misunderstood mission from an unvisited corner of the building called to mind a kind of mycological Fox Mulder. The impression was reinforced by a slight resemblance to '90s-era David Duchovny, and the poster in his office featuring a silhouetted mushroom floating UFO-like above a forest tree line. Underneath, it read: I WANT TO BOLETE.

Officially, the laboratory and all its equipment were available for any of the other museum departments to use, but it had been thoroughly colonized by the fungus folks. "In practice it's really my lab, because the other curators aren't doing genomics," said Dentinger. "They're sort of an older generation, by and large, and so they don't naturally do this work. But I've been pulling them into projects."

One of those projects involved "pulling in" Eric Rickart, the curator of vertebrate zoology, to test his dung samples. More specifically, dung samples from the animals he researched, which Dentinger's lab tested to assess their fungal content. Similar research into the northern flying squirrel, for example, suggested that as much as 90 percent of its diet was made of mushrooms.[7] This information helped to inform a hypothesis of Dentinger's, which suggests that, for the fungi, being eaten and moved by animals—as opposed to simply launching spores—might be the primary mode of reproduction over long time scales.

The mycorrhizal mushrooms he studied were often highly inbred, their spores mating with others from the same individual fungus, near the base of the same tree. When mating is found to have occurred between the spores of two *separate* individuals, genetically distinct and likely geographically more distant, it may be thanks to the ride provided by foraging critters—in this case, critters that also flew. Dentinger was quick to emphasize that the idea was mere speculation. "It's gone from speculation to slightly bullshit," chimed in his lab assistant, PhD student Keaton Tremble. "So it's moving up in the world."

Short of picking through squirrel scat with tweezers and a microscope, this research is only possible because of genetic sequencing and what is called polymerase chain reaction. In PCR, the basic idea is to hijack the machinery that DNA uses to target and copy a specific gene or section of genetic code.

Here's a very basic overview of how it works: You start with a small piece of fungal tissue, taken from deep inside the mushroom to minimize contamination by other microbes. Crushing the sample with a little mortar and pestle helps break open the hyphal cell walls and release the genetic material inside. From there, various bits of cellular debris must be separated from the nucleotides—the As, Cs, Ts, and Gs that define every living thing—using a series of solvents and spins in a centrifuge. After a few more filtering steps, what remains is a snotty mass of pure, concentrated DNA floating in the test tube, what they call "the squid," raw genetic material.

That material is mixed with primers that target the starting and ending points of the gene or section of DNA being sought, then run through a series of temperature changes that move the hijacked genetic machinery forward. By the end of the process, lots and lots of copies of the targeted section have been produced and can then be run through a device that reads with high precision the exact sequences they contain.

From there, the segment could be chosen for the kind of organism it identifies; fungi, for example, are often sought by targeting a region called the internal transcribed spacer, a string just a few hundred nucleotides long, basically functionless besides creating a gap between two active genes. Since the ITS isn't responsible for such things as building eyeballs or triggering an immune response, the contents of its code are more free to vary over time, making differences easier to track and compare. In general, closely related fungi will show fewer differences in these regions; distantly related fungi will show more. Compare enough of these tiny strings of nucleotides, and a magisterial picture begins to emerge, of cascading lineages of evolutionary inheritance, depicted in a relational tree called a phylogeny.

In Dentinger's lab, one such phylogeny hung from the wall, made from sheets of paper taped end-to-end. Down its length, densely clustered ladders of lines and numbers represented a computer's interpretation of the relationships between hundreds of sequences taken from just as many mushrooms, suggesting how they may have spread and evolved across the globe and over time. Color-coded text indicated which continent a specimen came from; closely related mushrooms naturally tended to group together in blocks of the same color. Yet some clusters included one or two lines of a distant continent's color, suggesting hitchhiking fungi on imported trees, or perhaps a closer-than-expected relationship between geographically distant mushrooms. Sometimes the phylogenies are oriented as nested circles, reading like rings in the tree of life.

As I struggled to make sense of the printout, Dentinger retrieved a small white box from the cabinet, something that would look at home on the shelves of an Apple Store. He removed from it a slender silver brick, not a smartphone but rather the latest in portable genetic sequencing, from a company called Oxford Nanopore Technologies, and it represented a step toward making the work they were doing in the lab possible in the field. "All of that stuff that we're talking about has always been part of the academic effort," said Dentinger, "and very few people have access to that outside of academia." The sleek brick he held in his hand suggested things were about to change.

Part of Dentinger's goal was to begin comprehensively document-ing the fungi of Utah and the wider intermountain West, an outsized project for a lone researcher and a handful of students. That was why, since signing on at NHMU, he had been coordinating with the local mycology clubs to help source fungi for the museum's collections. His lab handled all the sequencing, offering the club members an ID of their finds in exchange for their efforts. For the amateurs, the ability to do the sequencing itself would mark a real shift in that relationship, making it easier to identify their specimens and determine whether they represent new or interesting species of fungi before sending them in.

All the genetic and taxonomic information gathered by the lab, from its amateur contributors as well as from outings with Dentinger's stu-dents, was plugged into MyCoPortal, a global database used mainly by universities, botanical gardens, and museums. The entries are "vouchered," meaning that there's a physical specimen associated with the data, which other scientists can request for their own research. Dentinger pulled up the global entries, predictably opting for boletes. Strangely, certain geo-graphic details of South American porcini, a species of bolete, had been hidden. "I think they're concerned that people are going to go to that latitude and longitude to try to collect porcini," Dentinger hypothesized.

The ability of pickers to track locations of prime mushrooms in dif-ferent hemispheres could carry significant implications for their market value. But gaining a better sense of where fungi appear, and when, is also helpful for understanding them more broadly. Mycology, generally speaking, is in a phase of development that many other fields have long since moved beyond. Besides the relatively recent recognition of Fungi as a distinct kingdom, animals and plants are less ephemeral and easier to spot, and compared with fungi are quite thoroughly documented throughout the planet.[8] Among those working in other wings of the nat-ural sciences, said Dentinger, to spend time and energy on a fundamental inventory of organisms can seem a bit like stamp collecting.

"Unlike all of botany and virtually all of zoology, minus entomology, we don't have that information," he said. "We are in the midst of our own Victorian revolution when it comes to documenting fungal diversity; we've lacked that revolution, and I feel this professionally. It's very difficult to explain the importance of what I do to my colleagues—for example, in the Biology department—because they take this basic level of documentation for the organisms that they work with for granted."

The Extremophiles

After my first day at the NHMU, a spontaneous midnight outing with a couple of new friends took me to the southern shallows of Great Salt Lake (GSL). Behind us, a hare moon hung over the 1,215-foot stack of the Kennecott Garfield Smelter, its light suffusing the air in a glow that vanished across the lake, an atmospheric Rothko painting pinched at the sides by the silhouettes of Antelope and Stansbury Islands. My friends stripped off their clothes and ran giggling into the ankle-deep waters, disappearing into the murk while I stayed prudishly but contentedly ashore.

"Dim and pale, the moon, the ghost of a dead world, lifted above the distant Wasatch peaks and stared at the acrid waters of a dead sea," wrote artist Alfred Lambourne in 1909, describing an experience at GSL seemingly similar to my own.[9] But the brackish expanse is far from dead. As my friends scampered off into the darkness, their feet kicked up waters rife with phytoplankton, cyanobacteria, brine shrimp, and the larvae of brine flies. Also thriving in the water were a menagerie of halophilic (salt-loving) microorganisms including, of course, fungi, a fact only recently discovered.[10]

Two days later, I took an afternoon drive to the lake's northern section, easing my rental car over bumpy dirt roads en route to *Spiral Jetty*, the famous 1970 land sculpture by artist Robert Smithson. The whorl of black basalt stones was set upon a brittle shore, each step crunching on crystalline littoral, glittering neon pink thanks to its own populations of microbial life.[11]

Completion of a railroad causeway in 1959 bisected the lake into northern and southern arms, which have remained ecologically distinct ever since. The northern waters are of much higher salinity, sometimes as high as 34 percent, ten times that of ocean water. The southern half averages about half the salinity of the north and, unsurprisingly, plays host to greater biodiversity.[12] To survive in either environment, microbes must resist the moisture-sapping force of the salty water by absorbing osmotica, essentially anything that builds supportive pressure within their cells. What's more, they have to survive desiccating conditions in the evaporation-prone salt water, along with exposure to high levels of ultraviolet radiation, from constant exposure to sunlight throughout the day. Populations of these multi-extremophile microbes bloom and ebb in relation to the lake's water level, contributing to ever-shifting hues that are often described in mineral terms: emerald, sapphire, turquoise, cobalt.

In places along the shore, the salt formed a thin membrane that, when pressed, released tiny subcutaneous rivulets. Elsewhere the layer was firmly coagulated, and stooping down revealed the carcasses of grasshoppers entombed in pale pink lattices of halite and gypsum. At certain moments the glassy water resembled a mirage, and it seemed a vision straight from the pages of Ray Bradbury. The association was more than superficial. "If there was life on Mars, the last life that was there would have been salty," said Dr. Bonnie Baxter, a molecular biologist and professor at the Great Salt Lake Institute of Westminster College. Following our conversation, she took part in a NASA proposal to decide which sites on Mars they would next examine for signs of life. "We used to say there can't be life on Mars, because there's not enough surface water, or there's too much UV radiation, or all the lakes have dried up. But my microbes live in dried-up places with high UV radiation, they can survive for years among the salt crystals. So we think Great Salt Lake is an excellent analogue."

I had heard from Dentinger about the extremophile fungi in GSL, which led me to call Baxter after my visits to the lake. Baxter had spent years documenting the varieties of microbial life that thrive in GSL's waters, along with the scientific history of the lake, particularly after occupation of the greater Bonneville basin by white settlers in the mid-nineteenth century. In 2019, Baxter, whose scientific research focused mostly on the mechanisms of DNA repair, also copublished the first documentation of thirty-two species of fungi that call GSL home.[13] The fungi were isolated by Baxter's colleague, Polona Zalar, a biologist at the University of Lju-bljana in Slovenia, who collected samples from GSL while visiting Utah for a conference. The fungi's functions in the lake's ecology remained a mystery. For over a century, GSL has been the subject of biological study into the copious bacterial, insect, animal, plant, and other life that call it home. That's not because they just arrived, though, but because they were difficult to detect without genetic methods such as PCR, which wasn't invented until 1983. Up to that point, many species of bacteria and other microbes had been documented in the salty waters decades prior.

Upon taking her current role in 1998, Dr. Baxter was on the lookout for subjects that her students could engage with, and naturally figured the massive, unusual body of water just a few miles away fit the bill. "I thought that I would just, you know, find the person who was studying microbes that lived in high salinity—because there must be *somebody*—and I would ask if I could use those microbes and do my DNA-repair studies. But it turned out nobody was studying the microbes of the lake." One might think the largest lake west of the Mississippi, host to such an unusual ecosystem, would feature near the top of many biologists' lists of interests. "If you were to look at Lake Erie biology, you would find that the microbiome is really well characterized," said Baxter. "The fact that nobody knew what was going on here was pretty weird." It took international collaboration on a side project, as Baxter described it, to conduct a concentrated search for fungi in GSL.

At the outset of her research, Baxter met with the late microbiologist Fred Post, who had retired from studying microbial life in the northern part of the lake through the 1970s and '80s, which he'd pursued precisely because of the dearth of such research at the time.[14] Post's studies focused on archaea, bacteria, and the biological processes that led to unique phenomena such as gas domes, birdbath-sized geysers on the lake floor that become visible when the waters recede. Regrettably, by the time Baxter met him, Post had lost all his model organisms in a disastrous freezer-cleaning accident. Unable to share his collections from the lake, he encouraged Baxter to start making her own, and that's just what she did. Along the way, she began compiling the fascinating history of microbial science in GSL. "I found to my surprise that there were a lot of people that had done some early microbiology on the lake, and there were a number of women who had done this early microbiology, and this work was not published and it was kind of buried in people's theses somewhere," Baxter recalled. "I had to do a lot of digging."

Great Salt Lake is part of the Great Basin, the largest contiguous inland watershed in North America. Around thirty thousand years ago, the sprawling depression contained Lake Bonneville, estimated to have covered nearly twenty thousand square miles, including contemporary western Utah along with parts of Nevada and Idaho. Artifacts suggest that human presence near Lake Bonneville's shores extends at least as far back as 14,500 years, likely moving with the water level as evaporation and geological shifts shrank them to GSL's present shape and size around 13,000 years ago.[15]

Even after such a steep diminishment—or perhaps because of it—the GSL remains the largest saline lake in the Western Hemisphere. It's a major stop on the Pacific Flyway, where some 250 species of waterbirds visit by the millions each year. American avocets, Wilson's phalaropes, black-necked stilts, marbled godwits, and various other migrators take rest stops at this salty oasis to feed on the flies and brine shrimp living in and around GSL's waters.[16] In turn, they feed on a diverse microbial ecosystem of algae and other aquatic microorganisms, unknown until quite recently, that fill ecological niches at various depths. As the area eventually became a hotbed of mineral extraction—sodium chloride for deicing roads, magnesium chloride for steel production, potassium sulfate for fertilizer—a number of evaporation ponds were carved out at various points around the lake.[17] Each pond supports distinct microbial populations, and therefore each presents a unique hue, looking like watercolor swatches when seen from above.[18]

In its freshwater days, the larger, ancestor lake served as a rich fishing resource.[19] As the water level dropped and salinity rose, it may have emerged as the center of a local salt trade.[20] But like the fungi that still live there, much of the previous human activity in the region remains cryptic. "It's pretty hard to get that evidence," said Baxter, "because salt dissolves in water." What is known is that until a thousand years ago, the Shoshone and Ute peoples lived on the northern part of the lake, the

Gosiute to the south.[21] These communities lived and worked in intimate connection with GSL, and knew it in ways that science doesn't often prioritize—for example, by smell.[22] "They described the microbiology before they understood what microbiology was," said Baxter. "Understanding the smell, or seeing colors, or just observation was actually kind of early science, without the microscope."

In 1847, the Overland Stagecoach and the Pony Express brought Brigham Young and his fellow Church of Latter Day Saints travelers to GSL.[23] A familiar story played out, in which the Gosiute were displaced by encroaching militias, and eventually the US Army itself, culminating with their compulsion to sign a treaty in 1863 that pushed them out of the Great Salt Lake Desert entirely.[24] The fur trade brought men like Jim Bridger, who claimed first discovery of the lake and revealed his deep sense for the place by announcing that he'd reached the shores of the Pacific Ocean.[25] Eventually, the newcomers got their bearings, and the lake quickly became a site of recreation. To the newly arrived faithful, it represented a natural, pungent metaphor for the Dead Sea. Thus began our modern understanding of life in GSL, treated by its new occupants as a cultural blank slate.

"Silence, how dead! and darkness, how profound!" wrote Howard Stansbury of the US Corps of Topographical Engineers in 1847, quoting the poet Edward Young. "Nor eye, nor listening ear, an object finds."[26] Stansbury conducted one of the earliest federal surveys of the landscape surrounding GSL in 1849. He and his team documented diverse animal life around the lake, yet Stansbury remained so confident of the water's lack of life that he ordered his men use it to salt their supply of meat. Lucky for all involved, the briny microbes were benign enough that they did little harm besides adding a fetid taste; this putrid seasoning was given the nickname "salt junk."[27]

Since then, study of GSL's microbiology has carried forward in fits and starts. Alpheus Spring Packard Jr., part of the 1870s Ferdinand Vandeveer Hayden expedition, studied the geology and biology of the western US territories. Packard sampled algae from the lake, sending them back East to botanist William Gilson Farlow, who first isolated and identified cyanobacteria from them in 1871. The pink color of the northern waters is a result of microbes filled with pigmented carotenoids, the same molecules that give color not only to carrots but also many other living things, such as fish, flamingos, tree leaves, and algae.[28]

In the 1890s, biologist Josephine Tilden conducted what may be the first systematic study of GSL's microbiology. A leading algologist who also happened to be the first female professor at the University of Minnesota, Tilden was already a prolific collector of Pacific algae from around the world, the samples of which were kept with the plant collection in the U of M herbarium. She later absconded to Florida with many of the specimens.[29] Her ventures west landed her at such inland hotbeds for extremophilic life as Yellowstone and, eventually, GSL. The five examples of algae she isolated there, as well as the environmental descriptions she recorded, laid significant groundwork for future studies.[30] But in addition to her significant research contributions, Tilden has come to represent a pioneering figure for many female scientists. "If I could ever sit down and have a drink with somebody, it might be her," Baxter said of Tilden, whom she also described as "a badass." Photographs from the GSL excursions show Tilden in a long dress, surrounded by a group of gruff bearded men. In the context of early-twentieth-century science, a field that was even more male dominated then than it is today, Tilden seems like the toughest of the bunch.[31]

Decades later, in the mid-1930s, biologist Ruth Patrick discovered some of the same algae species around the lake, in its waters and in nearby soil samples. She conducted this work as an unpaid volunteer for the Academy of Natural Sciences, but eventually became a paid employee and a leading expert in diatoms, the group of algae that she and Tilden found abundantly in their surveys of GSL. Patrick's studies, though, were the most complete. Some of what she found were determined to be the remains of ancient algae; others were deposited by inflows of fresh water; some were native extremophiles that could handle the high levels of salt.[32]

Around the same time as Patrick was conducting her studies, at the University of Utah, Ruth Kirkpatrick (similar name, different person) was shifting her focus from an undergraduate degree in engineering —her yearbook photo from that time describes Kirkpatrick as "the Engineer Queen"—to a graduate degree in biology.[33] Her master's thesis consisted of a review of GSL microbiology literature, and she was the first to collect samples from the lake that were marked for both location and time. In addition to producing vivid illustrations of the life she collected, Kirkpatrick advanced approaches for culturing the communities of extremophilic algae in the lab, to figure out which lived in the lake itself and which had been deposited by one of several freshwater inlets. Within a couple of decades, the University of Utah gradually drew down its studies of GSL microbiology. By 1961, the rail causeway was built, forever transforming the lake's ecosystem, and study into the life it harbored mostly stalled until Fred Post took it up again in the '70s.[34]

In all that time, no one ever found fungi. At least, not knowingly. Around the same time as Kirkpatrick and Patrick were conducting their studies, University of Utah graduate student Winslow Smith and marine microbiologist Claude ZoBell collaborated on a technique that involved dunking microscope slides into GSL's waters, so that they could document the communities of microorganisms that grew on them. At one point, Smith dunked a device to test for sterile conditions in the lake, which emerged totally encrusted in salt crystals and grew covered with mold. Smith threw the slides out, deeming them contaminated, but in doing so he might have missed the first-ever evidence of fungi in GSL.[35]

Now that fungi are known to dwell in GSL, it's easy to hear stories like these and wonder how early scientists could have missed what today seems so obvious. But microorganisms are anything but intuitive; until recently, hypersaline environments were investigated with a focus on bacteria and archaea. Since then, halophile researchers have isolated yeasts and filamentous fungi in other salty lakes, including the Dead Sea.[36] New and unknown species have since been found, too, including in wood and plants near salty shores, in microbial mats and brine ponds.[37] With DNA sequencing, it's possible to find signs of life that simply eluded the microscopes of earlier scientists.[38]

But those techniques have already been in use for almost fifty years. It wasn't until the twenty-first century, when Baxter and her international colleagues thought to look for fungi in GSL, that a thriving community of them was found in its waters.[39] Baxter credits that largely to the difference in perspective that her international colleagues brought to the project. "The funniest thing I heard them say was, 'We study the things that you guys throw away,'" Baxter recalled. "Meaning that if I'm pulling halophilic archaea or bacteria out of Great Salt Lake and my plates get fungi on them, I throw them in the trash, and I say, 'Oh, they got contaminated,' instead of saying, 'Oh, that is also on the growing media, and also must be a halophile.' And so these guys say that they study our contaminants, which is hilarious."

———————

There was one place in particular I'd hoped Dentinger would show me before I left Utah. Down the hall from the NHMU genetics lab, he tapped a security card against the sensor beside a pair of tall doors, and I followed him into a high-ceilinged warehouse occupied by row upon row of large, long cabinets set on rails. At the end of each movable corridor were catalog numbers, bar codes, and pictures to help navigate the 135,000 specimens they contained. Turning down an aisle near the middle of the array, we walked with echoing footsteps to the far end, where there sat a lone, unremarkable white cabinet. "There it is," said Dentinger with a note of ironic understatement. "The fungarium."

All the museum's fungi, just a few dozen specimens, the first of them gathered in 2017, took up the equivalent of a small kitchen pantry. It was a modest collection that represented a great deal of progress. "There was nothing here when I arrived, not a single specimen," he explained. "There was not a mycology collection here, so I started one from scratch." Dentinger was actually the institution's very first mycologist, and knew of only two other collections in the entire state: one at Utah State University, where the staff mycologist had retired and was never replaced; and another at Brigham Young University, which focused on lichens. It was down to a small team—basically, him and his students—to do all the field collecting, hence, this fledgling fungarium was burnished by the contributions of outside amateur groups and local mycological clubs. "I'm working to try to get them integrated into the process of scientific collection and donating to the museum," he said.

Mycology has long been built largely on contributions from communities of amateurs and enthusiasts.[40] After all, until the second half of the twentieth century, being a professional mycologist essentially meant working in a basement office of the botany department.[41] Getting people to pick and identify mushrooms necessarily meant connecting with people who undertook a study of fungi for its own sake. But as fungi have grown increasingly popular, along with excitement about the possibilities they pose, they have seemingly attracted more people to pursue academic study. Dentinger reported that his classes are noticeably more full than they used to be, populated with many students attracted to some degree by the growing acceptance of the claim that fungi can "save the world," fruits of a rising trend of myco-evangelism that has popped up in TED talks and fringy fungus festivals and is reaching ever more mainstream audiences.

"I think there's a lot that that end of mycology can contribute to the future of mycology as a whole, but it will require some more empiricism," Dentinger told me, choosing his words carefully. "And the other way around, I think it injects a lot of creativity and imagination that can really propel the field forward in ways that are unexpected. Hypotheses have to be generated somewhere. But it's the testing of those hypotheses where you distinguish real science from hand-waving and storytelling."

Science can seem synonymous with being advanced education and laserlike focus on a hyperspecialized corner of research. That increasingly describes the work of many professional scientists, especially as DNA-focused techniques take center stage. But there is also plenty of room for contribution from outside, even just by picking an interesting mushroom. Mycology, like any natural science, relies on the collection of physical reference specimens, kept for posterity in collections such as the closet-sized fungarium at NHMU.

But science funding has long been thinly spread, and grows thinner each year.[42] The institutions where collections are gathered and preserved risk losing the resources to build and maintain them, meaning that what information does exist may languish as departments reshuffle and the lack of job prospects induces fewer people to follow the path that might lead them into a role where they can carry on the work of building the fungal record and deepening science's understanding of the kingdom. Fungi have always been a bit fringe, even among biologists and naturalists, and in some ways it's not all that different today. What is different is that this fringe might keep the broader field alive over the coming years, within and outside the traditional institutions.

CHAPTER FOUR

Big S, Small S

I n the South West of London, the Royal Botanic Gardens, Kew, are
bound by the contour of the River Thames to the north, high walls
everywhere else. A storied institution of natural sciences, Kew is home
to more than eight million specimens of plants and fungi.[1] Its vast lawns
are populated by diverse, neatly spaced specimens of flowers, shrubs, and
trees, with stately greenhouses and courtyards presenting the very image of
an imperial conservatory, a grand sight even in the gray winter.

Pacing the length of the southern wall, I found the side entrance
and the big yellow security phone. Hanging it up, the gate swung open
to reveal the Jodrell Laboratory, where Dr. Brian Douglas waited for
me outside. With a soft-spoken and easygoing manner that contrasted
with the magisterial environment of Kew, he led us inside to an atrium
where big windows separated us from a gleaming two-story laboratory.
We chatted for a moment as, on the other side of the glass, scientists
bustled busily about in the bright white space. After a quick overview of
the Mycology department's work, we headed downstairs, passing a series
of large, vivid color portraits of pollen and descending a steep staircase
that led straight to the door of the fungarium.

Tall concrete-walled corridors were lined by aisle upon movable aisle
of pea-green storage boxes, stacked a dozen high and stretching toward
the vanishing point. All told, they contained some 1.25 million different
species.[2] More than fifty thousand of them were type specimens, those

that mark the basis of an organism's formal recognition by science. It is the largest reference collection of fungi in the world. "Think of this as a Rosetta stone for fungi," said Douglas. Sequencing the entire fungarium collection, he said, would be a benefit to fungal taxonomy comparable to what deciphering Egyptian hieroglyphics meant to archaeology.

Tours are common at Kew, so a number of the most crowd-pleasing specimens were kept out on display. Atop a glass case in the main throughway was a ram's skull, its horns sheathed in the chalky myce-lium of a bone-eating ascomycete. Next to it was a tub of Marmite—a by-product of beer brewing, the favorite English spread is derived from a fungus, every brewer and baker's favorite yeast, *Saccharomyces cerevisiae*.[3] Inside the case were many other fungal curiosities, including desiccated little orange globs, *Cyttaria darwinii*, or Darwin's fungus, brought back with the famous naturalist himself from Tierra del Fuego on the voyage of the HMS *Beagle*. His fading signature was scrawled on a slip of paper right beside them.

Much of Kew's massive collections comes from decades—centuries, really—of international scientific collaborations and collection efforts. But it also reflects a history of colonialism, a fact that is not lost on Kew and its leadership. "For hundreds of years, rich countries in the north have exploited natural resources and human knowledge in the south," wrote Alexandre Antonelli, Kew's director of Science. "Colonial botanists would embark on dangerous expeditions in the name of science but were ultimately tasked with finding economically profitable plants. Much of Kew's work in the nineteenth century focused on the movement of such plants around the British Empire, which means we too have a legacy that is deeply rooted in colonialism."[4]

Douglas was happy to indulge my curiosity as we peered into boxes and behind glass cases, but he was not there as a tour guide. Two months shy of finishing his contract as the community fungus survey leader for Kew's Lost and Found Fungi project, or LAFF, he described the five-year effort as having a threefold mission: to update the understanding and conservation status of fungal species in the United Kingdom, to make a lasting contribution to U.K. mycology, and to improve fungal education and awareness in general.

The first part involved running down a list of one hundred fungal species, about which there were few enough records to justify a reassessment of rarity. For a species to get on the LAFF list meant it had been found in fewer than five sites across the U.K. over the previous fifty years, or that it was newly discovered, or recently disambiguated from a group of similar fungi, among other criteria. Those deemed to be genuinely under threat would be red listed.

National Red Lists work like regionally focused versions of the International Union for Conservation of Nature (IUCN) Global Red List, both representing authoritative records of the species most at risk of extinction, and effective tools for mobilizing changes in conservation policy.

To get a sense of the disparity in conservation information for fungi, we can compare them to plants. As of this writing, in Great Britain, the official Red List includes 37 species of fungi, all of them boletes, compared with some 1,756 species of vascular plants. The pattern is also reflected on the global scale; as of March 2020, the global Red List contained 343 entries for all fungi compared with 43,557 entries for all plants.[5] Lack of information about the diversity and distribution of fungi, though, is a problem by no means limited to the U.K.

Fungi are generally underreported, in large part because they're simply not as easy to document as plants or animals. Fungi in the wild can be fickle, appearing and disappearing at varying intervals depending on environmental conditions, time of year, and other factors yet to be fully understood. The fungi that produce mushrooms are a small minority of the world's species, and they don't (one hopes) often make themselves known to observers by flying overhead or scurrying up a tree. Assessment of fungal species' conservation priority is therefore often qualified by the difficulties in comprehensively documenting them. Those listed as safe might in fact be threatened, but overrepresented in the collections; those listed as extinct might actually be plentiful, just underreported. Douglas also noted the inadequate documentation of habitat loss over the last fifty years, another key factor in understanding the threat posed to fungi, along with all other forms of life. The goal of the LAFF was to work toward "ground-truthing" the rarity of fungal species, as Douglas put it, and to get a more accurate sense of any threats to their populations. It was also to show how doing so was possible with dwindling resources.

Even an institution as august as RBG Kew faces budgetary limits that prohibit deploying the army of skilled collectors required to fill the gaps in the fungal record. In 2015, funding cuts left Kew with a budget deficit of some £5.5 million per year.[6] That meant projects like LAFF, no matter how essential or successful, had to be sustained through onetime, outside grants. But despite a strain of cultural disdain toward fungi, a robust coalition of amateurs and enthusiasts made LAFF possible, supported in their own "treasure hunts" for target fungi that then fed back into Kew's own reporting of regional diversity, and eventually to the Red List. "We're tapping back into the citizen science community, which we've always traditionally worked with to an extent," Douglas said as we settled in at a computer terminal to review the results.

The LAFF released a rolling list of species that amateurs could seek out and contribute. "Puffballs, brackets, smut and rust fungi, tiny cup fungi, leaf-spot parasites, lichenised fungi, and more," read the project website.[7] Some species were chosen because they hadn't been recorded in the U.K. for more than fifty years; others were known from only a few sites, so their relative rarity or abundance was unclear. Others were new to science, while still others had already been listed as worthy of conservation concern. A diversity of habitats was represented as well, "from dune systems to mountain plateaus, bogs, calcareous marshes, ancient woodland, orchards, or urban areas and gardens."

Nearing the end of its five-year life, the project had added some 1,400 new records of rarely recorded species; around half as many as had been collected in the previous fifty years.[8] As the project wrapped up, they had found seventy-seven out of their one hundred target specimens, spanning more than seven hundred different sites across the U.K. "Some species have gone from two or three records in the world to us knowing that they're widespread," Douglas said with a note of pride. "Some species have gone from us knowing they've recently arrived to the U.K., and now have spread; others we haven't found, so they're possibly still very rare, but you can't confirm that by *not* finding something." Five years is hardly enough to mount a comprehensive survey of an entire country's fungi, though, and especially with financial resources for taxonomy dwindling, Kew was limited in its ability to do much of any field survey work.

Traditionally, taxonomy involved a group of experts using their highly specialized knowledge of their local environments and the fungi that live there to identify new and interesting species, often based on subtle physical traits. DNA sequencing has added a whole new dimension to taxonomy, but still relies on physical reference specimens to make the genetic bar codes meaningful. Douglas reported that such "morphotaxonomy" was receiving steadily less funding, with almost none going toward conservation save for a few small grants from a handful of national organizations. As a result, when some interesting new mushroom comes in the door, there are fewer people to work out where it fits in the fungal menagerie. So, Douglas and his fellow mycologists turned to the amateurs to help fill the gap.

"I've been all over the U.K., on a variety of different buses, overnight trains up to Scotland for forays, workshops, and study weeks," Douglas said, noting that he doesn't drive. We resumed our stroll through the wide-open grounds of Kew, chatting as swans stalked us to the door of the cafeteria. "We realized that if we want to get results, we have to help other people, and establish goodwill for common shared aims, and that is what we do. We try to help the field mycologist community, which involves meeting in person, lots of emails back and forth, giving them lots of ideas, and encouragement, more than anything else."

Many of those field mycologists were part of independent regional clubs associated with the British Mycological Society (BMS). Founded in 1896, the BMS is the second-oldest national mycological society in the world. (France established the first, in 1884.[9]) The various groups affiliated with the BMS organize forays throughout England, Scotland, and Wales, representing a latent base of enthusiasm and knowledge that Kew could tap in its efforts to burnish the understanding of fungi in the United Kingdom.

Toward the end of the LAFF project, some of these groups were equipped by Kew directly to prepare their own DNA sequences, by way of laptop-sized Bento Labs; portable DNA extraction and amplification kits, each featuring an onboard centrifuge, thermocycler, and gel electrophoresis (a readout for visually assessing the quality of the results). Forayers could fetch fungi from LAFF's lists, or any other interesting finds, prepare the specimens for sequencing with the kit, then send them off for analysis. The hope was to better identify conservation-worthy fungi, but also to "upskill" the U.K. mycological community, opening the door for amateur field mycologists to submit DNA bar-coded specimens straight into Kew, or other international research institutions. This approach bypassed the bottlenecks created by a shrinking group of taxonomists faced with a backlog of new, as yet unidentified, specimens. The amateur field mycologist, taking all the steps themselves, could get information back about a sequenced specimen in three days for four pounds, plus postage, and the institutions got useful data on new and interesting species.

"The main difference with the field community is that they have the time to learn about general diversity whereas we need to specialize on certain groups," said Ester Gaya, senior research leader of the Comparative Plant and Fungal Biology at Kew. She joined us for a brief pit stop at the break room. "You need a lot of time to dwell on a group, you can't afford to do it all."

Historically, pretty much anyone could send a letter or, more recently, an email to a department at Kew to get answers about plants or fungi: What is this thing called? Have I found a new species? "All the way through the history of Kew, the mycologists have worked with amateur experts and enthusiasts," said Douglas. "That's happened from the 1900s all the way up to very recently. Where do you think most of the fungarium came from? Much of it is the collections of the amateur experts, and these collections have all been brought together in one place as a national reference collection. But now there's no funding for taxonomy, and there's no taxonomy training even close to the same extent as you used to have."

The dawn of genetic sequencing has represented a sunset of sorts for traditional taxonomy, the study of the features of organisms and how they relate based on their physical, observable traits. But those traits and the skills to identify them still play a crucial role. You can compare all the DNA sequences you like, but without specific sequences pegged to physical specimens, you can't really be sure of what you have. At Kew, there were two morphotaxonomists for non-lichenized fungi remaining—Paul Cannon and Martyn Ainsworth—already well into their careers, working part-time and with no clear successors lined up to take over once they retired. Traditionally, taxonomists would commit decades of their lives to their specialized knowledge, passing it on in a mentor-teacher arrangement. "You had an assistant mycologist and a head mycologist," Douglas explained. "Sort of like the Dark Lords of the Sith, but not evil."

The gradual decline of that tradition has left the question of who will carry the knowledge forward, and how. In the meantime Kew's mycological ranks have actually grown, and diversified their focus, embracing genetic sequencing, building out one of the only master's courses in the country involving fungal taxonomy, studying the impacts of pollution on ectomycorrhizae, even publishing a massive, first-of-its-kind "State of the World's Fungi" report in 2018. The three most recently appointed mycologists in senior positions at Kew were also women, including Gaya. "If anything, I feel mycology at Kew has revived extraordinarily in the last years," said Gaya, whose ongoing projects include leading the institution's involvement in the Darwin Tree of Life Project, which aims to sequence all fungi in the British Isles, also involving partnerships with the amateur field mycologists. "We mycologists are just evolving, the same way our subject of study does."

The role of an institution like Kew can be described as one of accruing, retaining, and dispensing biological knowledge, backed by physical collections, a cornerstone of taxonomy. Douglas described his own perspective on Kew's role as one of supporting the U.K.'s capacity for fungal taxonomy and knowledge. Given its reputation and resources, it acts as a vault as well as a gatekeeper. Kew's work with the communities of amateur mycophiles may serve as an example of how large institutions can adapt as funding for science dwindles, as the roles for specialized taxonomists become more scarce, and as scientific expertise grows more generalized and widespread.

DNA sequencing has gradually taken its place as the state of the art in natural sciences, and advances in technology have made it easier for nonprofessionals to do the basic work of documenting what fungi exist and where. "It is a lot easier to teach a field mycologist to do DNA work than it is to teach field mycology to someone who's got molecular lab skills," said Douglas. "One is essentially making a complicated cake, and then doing some pattern matching, and the other is teaching them about the biodiversity of seventeen thousand species in one country."

As Kew distributed its five Bento Lab boxes to various amateur groups around the U.K., it was with safe and easy-to-learn preparation methods adapted for fungi by Dr. David Harries, a dedicated amateur mycologist and cofounder of the Pembrokeshire Fungus Recording Network, which operates a small genetic sequencing lab out of the corner of his garage.[10] According to Douglas, the sequences he receives from amateurs are often on par in quality to those prepared at Kew, a testament to the care the enthusiast groups put into their work, and suggesting a real future for the nonacademic mycological community in generating data of high quality, and in significant quantities.

"It will be a full cycle back to the old days, where the people describing the fungi are the amateur enthusiasts, only this time they'll be DNA-enabled," Douglas said. "There's a lack of expertise at the top in terms of professional mycologists, so the knowledge pyramid has flattened, but it's got a hell of a lot wider."

———

After a day at the fungarium, Douglas took me to a pub around the corner from the lab, aptly named the Botanist. Soon we were joined by Nathan Smith, a young PhD student who had just that day begun a stint at Kew, bringing with him a keen interest in the history of British mycology. Between sips of beer and mouthfuls of chips, I learned about the foundations of British mycology.

The Victorian era was a boon for the natural sciences in Britain. The middle of the nineteenth century was the era of the naturalist clubs, such as the Woolhope Naturalists' Field Club, which incubated a culture of mycology that lives on today, organizing an early "foray among the funguses" that became an annual event on the first week of October.[11] Members would traipse about the woods gathering mushrooms into baskets, displaying their finds on a table, and deducing their identities while sipping tea, an agenda that any member of a mycological society in North America could recognize.[12]

"It becomes very fashionable," said Smith, himself a member of the regional mycology club in Cambridge. Mycology started gaining traction in nineteenth-century England, he said, thanks largely to the work of amateurs with broad interest in natural sciences. "You have a big rise in natural history societies, you have a big rise of field clubs, which are different because they're about going out into the field. They're also less elitist, so you have this big rise of field clubs that are less elitist; you have your weekends starting to be codified into law, so more people have free time; and you have the fact that by then no one's actually studied fungi." Of course, "less elitist" in Victorian England may be something of a skewed standard; the amateur mycology movement was led largely by country clergymen and doctors.[13] But even so, compared with the cloistered halls of the academic institutions of the time, an open invitation to wander the woods and look for mushrooms must have seemed like the realm of the hoi polloi.

To the Victorian naturalist, fungi represented a new field in which to make discoveries, something that was quite in vogue at the time. With fungi, though, it was possible to make discoveries in the backyard. "The rich folks did all the plants and animals of the U.K.," said Smith. "If you want to be a botanist discovering things [back then] you need to go out to the colonies. If you want to be a mycologist discovering things, you need to move five meters that way." Not unlike the early twenty-first century, availability of new technologies like the microscope boosted mycological studies in a context of rich naturalist tradition, to which Kew was central. Naturalist clubs of various sorts formed and enjoyed experimenting with the new tools of inquiry.[14] Mycology, marginal then as now, was near the center of the Victorian era of British naturalism that would also take root in North America.[15]

England was the birthplace of the Linnean Society, the world's oldest active biological society.[16] It is also home to the Royal Society, the world's oldest national scientific society, at one time presided over by none other than Sir Isaac Newton.[17] The Royal Society's history was intimately tied with the formation and ongoing funding of Kew itself. Joseph Banks, prior to presiding over the Royal Society, was part of the exploratory voyage of James Cook (also a project of the Royal Society), becoming the first "unofficial" director of Kew in its early form.[18] Miles Joseph Berkeley, widely considered the founder of British mycology, described around six hundred new species, and built up a collection of some ten thousand specimens, which he ultimately donated to Kew, helping to usher its arrival as a formal field of study.[19] The role of mycologist was at the time known as the Herbarium Principal Assistant of Cryptogams. The first time Britain actually named mycology an official field of study was in 1918, at the end of the First World War, when the Board of Agriculture created an advisory service that listed regional plant pathologists as "advisory mycologists."

Mordecai Cubitt Cooke was a colleague of Berkeley's, joining Kew in 1880 as a "cryptogamic botanist" (today, again, we would say mycologist).[20] An eccentric science writer with a coarse beard and a penchant for bowler hats and pipe tobacco, he had some problematic ideas about the connections between race and use of certain drugs, and was in many ways an outsider among his well-to-do colleagues. Mostly self-taught, yet firmly in the center of the world of fungi in Britain, he actively organized amateur clubs around England, including the first foray among the funguses, which gave rise to the British Mycological Society.[21] Cooke's successor at Kew, George Massee, was also involved in the founding of the BMS, as its first president, and was the last mycologist at Kew without any formal training in the subject, being a new area of inquiry, formal training was not to be expected. Famously arrogant, renowned scientist and historian of the era John Ramsbottom said of Massee, "If he had any capacity whatever for taking pains he would have been a genius."[22]

One especially notable figure who dealt with Massee was none other than Beatrix Potter. The author of *The Tale of Peter Rabbit* is less well known for her significant contributions as a mycologist and nature illustrator. Prior to her authorial fame, Potter spent a great deal of time in Kew's herbarium, producing hundreds of gorgeous illustrations of fungi, among other organisms, even presenting a paper, "On the Germination of the Spores of Angaricineae," to the Linnaean Society in 1897; at the time, women were all but forbidden from pursuing a proper scientific education, and were prohibited from scientific societies, so her uncle delivered the paper in her stead.[23] In that context, Potter's efforts at contributing to the sciences at Kew were, perhaps predictably, met with resistance. She appealed to Massee to review her drawings and writings on fungi, and although he "took objection to my slides," as Potter later wrote, Massee did submit her paper. It was unfortunately lost and never published, and the society issued a posthumous apology to Potter for the sexism she experienced.[24]

In mycology as in every other field, female contributions in and outside academia are many, if often overlooked.[25] But that situation has shifted in the years since; for example there are now more full-time female mycologists at Kew than males, and more mycologists on staff in general than ever before. Even in the Victorian era, male dominance of mycology was not total. After Massee's departure from Kew in 1915, his successor as head of Mycology was Elsie Wakefield, who went on to publish nearly one hundred articles on fungi, including the foundational description of *Psilocybe cyanescens* in 1946.[26]

With mycology now firmly established as an academic field of study, even as taxonomic and ecological research gives way to an emphasis on genetic sequencing, amateur mycology has experienced something of a revival outside traditional institutions, including and maybe especially online. Mushroom Spotters UK, a particularly active online community of mycophiles, is growing fast, maintaining a Facebook page of more than thirty-five thousand members as of this writing, adding thousands every month. It's one example of what seems to be a general swell of interest in fungi throughout the United Kingdom and beyond.

"The presence of Facebook has been, I think, a very big catalyst in this," said Rich Wright, a young and affable independent mycologist from Bristol who joined our table as the second round of beers arrived. An active organizer of various community mycology groups, Wright had been brought in to work with Kew as something of a liaison with the amateur recording communities. He reported noticing a distinct spike in interest around the subject throughout the U.K. "It has exponentially exploded in the last four years."

Explosions aren't known for being orderly, though. Everyone at the table cited what they saw as growing pains in the putative online mushroom communities, fraught with exhibitionism and the signs of a culture still establishing its norms and customs. While information of real value was exchanged on these forums, they were also becoming venues for what Wright bemoaned as pseudoscience and bickering. "There isn't any way that you can cut into that without ending up in some ridiculous Facebook argument," he said.

Smith chimed in, "I've done it once or twice and I won't touch it again," adding that he thought the problems were mostly the signs of an adolescent phase for the community. Across the pond in the United States, social media also manages to be toxic around mycology, as with any other subject. They voiced concern about certain aspects of American mycological culture they'd observed making their way back from the USA. For example, the growing popularity of potentially damaging foraging practices, overheated assertions about mushrooms' uses or abilities, and the surge in home cultivation that might introduce non-native strains into British ecosystems. But they also noted a growing citizen science community in the United States that was spreading knowledge and skills and making real contributions that were redefining the field. In their work, they hoped, they might be helping to foster something quite similar in the U.K.

So You Want to Be a Mycologist

After crossing a rain-slicked street in downtown Brooklyn, I ducked under the awning of 33 Flatbush Avenue. The century-old former bank had gone decades since its last renovation, so the broken buzzer was hardly a surprise. As I dialed the number scrawled on a taped-up sheet of paper, someone on their way out let me sidle through the door, into the dim lobby and up the stairs. (The elevator was out of order, too.)

Nearing the fourth floor, I narrowly avoided colliding with Craig Trester as he rushed in the opposite direction. In addition to leading the night's mycology class at Biotech Without Borders, it was his task to fetch each new arrival from the front door. After a hurried greeting, we passed through a long, unlit hallway flanked by dark, messy workspaces and the half-visible fragments of artworks-in-progress. Under the ownership of an eccentric old-timer named Al Attara, the seven-story Metropolitan Exchange—MeX, for short—had since the late 1970s played host to

shifting casts of creatives, makers, citizen scientists, and offbeat entrepreneurs.[27] It was a space with sufficient purpose and community feeling
to have resisted the insistent force of gentrification. "This place is one of
the last little chunks of old New York," said Trester as we stepped into a
brightly lit, cement-floored room where the class was to be held.

Office chairs lined the length of a steel table, with a flat-screen
TV set across from a comfy sofa and an impressive communal library.
Despite the slight vibe of a basement den, it was really the classroom
space for a community biology lab. Farther on, a Biosafety Level 1 lab
space, certified to handle low-risk microbes, with a door that led to a
self-contained room rated for Biosafety Level 2, allowing for working
with "mammalian tissue cultures," as Trester explained it.

Atop nearly every surface sat glassware, centrifuges, freezers, thermocyclers, all bought at auction or donated. Just days after our class, Biotech
Without Borders would acquire a scanning electron microscope, hauled
in the back of a rented truck from a plant genetics lab on Long Island.
With limited resources, BWB had managed to bricolage an impressive
capacity for far more meaningful science projects than merely creating
multicolored flames from Bunsen burners or goading foam to erupt
from beakers. One of their favorite experiments, for example, involved
modifying the genes of algae to fluoresce under ultraviolet light.

"Before 2008, there were no community biology labs," said Trester,
introducing the facilities to our group as we settled in and shook off the
December drizzle. "On the back of the financial collapse, a lot of biotech
startup companies went belly-up, so all this biotechnology equipment
you see went on the auction market."

Trester's presentation for the evening was evocatively titled "Fungi
Are the Future: How Mushrooms Will Save Our World," a nod to the
popular TED talk by Paul Stamets, which was part of the inspiration
for Trester to pursue his path of fascination with fungi. The intention
of the class was to guide the small but enthusiastic group of neophytes
down the first steps of that same path. Earnest and energetic, Trester was
possessed of an enthusiasm for science that clearly took some effort to
contain, expressing itself in staccato signals of constant information as
he loped among scientific subjects and domains. The enthusiasm might

have been due in part to his self-avowed caffeine addiction, but fungi had also presented a uniquely exciting subject.

At Trester's behest, everyone around the table introduced themselves while he dashed downstairs once more. Among the eight attendees was a former publisher, the owner of an herbal medicine store, a vegan chef, a fashion designer. Almost everyone was in the early stages of their study of fungi. When Trester returned, he assumed his station next to the computer monitor, powered up the slideshow, and got under way. "I'm going to start with a presentation," he said in a cheery baritone, tugging unconsciously at his wiry beard. "It's going to have a lot of information."

The slideshow began with a photo of a chunk of asphalt, every centimeter of it threaded through by mycelia. Trester had taken the picture at a mycoremediation workshop at Van Cortlandt Park in the Bronx, and as of that moment, he recalled, "I became hooked and read as much material about the topic as I could find." Skipping around the two-hundred-odd slides in his presentation, he led us through a horizon-spanning survey of fungal biology, ecology, and evolution. In video shot through a high-magnification microscope, we watched hyphae branch and grow as illuminated nuclei coursed throughout their length. Another video demonstrated the fusing of intersecting hyphae, in a process called anastomosis. Our session would cover only a fraction of all the material Trester had organized. Every raised hand presented a question that he pursued with gusto, exploding any structure the class may have once had into an improvised tangle of tangents and sidebars.

Trester seemed at times on the verge of losing some people, as molecular diagrams of organic polymers flicked across the screen, but no one complained; the presentation, scheduled to last for three hours, stretched out to five. There was just so much to cover, and Trester was evidently still working out how best to organize it all for a mixed audience. To get everyone on the same page, he held a bag of mycelium-infused grains over his head at the outset and announced: "This is fungus." Then he cut open the top and passed it around the room, encouraging the room to take a whiff of the spawn, with its strangely pleasant, sweet aroma.

Over the preceding year, Trester, whose formal education is in history and political science, had succeeded in forging his interest for biology and biotechnology into an enthusiastic, multifarious career that had allowed for studying, practicing, and teaching the subjects that most fascinated him. Still in his twenties, he'd spent the previous three years working as a barista, bartender, and tutor of Chinese, until his scientific interests drove him to recognize a growing bioeconomy, in which he saw opportunity to support himself by educating people about fungi and the broader complexities of soil biology. Becoming an instructor with Biotech Without Borders and Genspace—the city's predominant community biology labs, both housed at MeX—he pursued his own mycological research in parallel. Trester had managed to sustain his budding career by way of a constantly growing personal network, even teaching sessions in the city's public schools with the support of the National Wildlife Federation and the Department of Education, developing mycology and soil ecology curricula for the city's magnet schools. "It's a pretty unique situation," he told me.

But Trester is not alone in plying passion for mycology into a career outside academia. Educational workshops like the one he taught were staples of an emerging culture around fungi in North America, an informal initiation that most modern mycophiles undertake in some form or another. Upon mastering the fungal foundations in these courses, many, like Trester, go on to teach their own versions. No matter where you catch one of these informational boot camps, or from whom, they often strike many of the same notes: the basic structure and life cycles of fungi, their ecological roles, their many uses and cultural associations. There is no official fungal pedagogy in this emerging world of citizen mycology, but rather a common vernacular aimed at familiarization with fungi and their many facets, the basics of growing mushrooms for food and medicine, taxonomy and identification, cloning and tissue culture, experiments with soil remediation and biomaterials.

Visual, tangible, even sniffable aids are standard fare. As the young educator spoke, his audience passed around mushrooms that had been strewn about the tabletop. Oftentimes, spawn or grow bags, petri dishes or jars containing fungal cultures, anything that can help to connect the information to the senses will also make the rounds.

For this course, the hands-on portion also took the form of cardboard strips, to be inoculated with the mycelium of oyster mushrooms. Trester juggled spritzers filled with rubbing alcohol, fungal spawn, strips of wet cardboard, and a piping-hot kettle. "Don't try this at home," he joked. "I used to be a barista." After a spray-bottle sterilization of hands and tabletop with isopropyl alcohol, students briefly soaked their rectangular scraps of cardboard in water, layered them with crumbled oyster mushroom spawn, and rolled them into little fungus-filled burritos.

Everyone was encouraged to return home with their spring roll–sized tubes, sweaty inside their plastic sandwich bags, to watch in the coming days as they became carpeted with fluffy white mycelia, hungry for more substrate. What would be done with the burritos after that was an open question. They might become the start of an at-home cardboard-composting project, digesting the cardboard or other refuse and eventually giving birth to edible mushrooms, assuming the absence of potentially toxic ink. With some patience and care, the inoculated wraps could be worked up into the basis of a small mushroom farm. Most, though, would probably end up tossed away, in which case, they represented at least some small amount of mitigated waste as the fungus digested the cardboard. No matter what, the students had made contact with a vast kingdom of life that many never experience beyond the borders of their dinner plate. Like the burritos themselves, the hope was that the students would become vectors for fungi, carrying on and sharing their newfound knowledge of and curiosity for the subject. It was all part of a process of demystification that each person could carry on in their own networks, to spread the proverbial spores.

"The cool thing about science is basically that you should be able to reproduce any good experiment," said Trester to the class. "And that's kind of the big goal here, to give you the access, and the tools to do so."

———

For weeks I had plans to be in Oakland for a community DNA-sequencing class with citizen scientist célèbre Alan Rockefeller, but the COVID-19 pandemic canceled them. Instead, just like nearly every other social event as of March 2020, the workshop would be held via live video stream. From a friend's living room, I logged on with more than a hundred others as Harte Singer, organizer of the event, introduced us through his smartphone to Rockefeller and the uncharacteristically quiet environment of Counter Culture Labs.

CCL is a community biohacking lab tucked into a corner of the Omni Commons, next door to a hacker space and just down the hall from an East Bay chapter of Food Not Bombs. Something like a West Coast version of Biotech Without Borders, it's home to a local community of tinkering mycologists, most notably Bay Area Applied Mycology, and driven by an open-source ethos, operating on volunteer labor and membership dues.[28] The place was stuffed with workspaces and hardware, some bought, some borrowed, much of it received through a nonprofit organization that redistributes equipment offloaded by the Bay Area biotech industry. Under a big banner reading CITIZENS FOR SCIENCE, SCIENCE FOR CITIZENS, bio-curious tinkerers gather for projects ranging from DIY insulin to vegan-cheese cultivation. "For far too long, science has been locked away in the 'ivory towers' of universities and research labs," reads CCL's website.

On our screens, Rockefeller, in his seemingly permanent ensemble of T-shirt, baseball cap, and short-kempt red stubble, juggled a series of sampling instruments. Attendees had originally been encouraged to bring their own mushrooms for sequencing, but since that was no longer an option, Rockefeller had resorted to mushrooms from his own outings. Extracting rice-grain-sized chunks of fungal flesh, he demonstrated with precise, practiced movements how to prepare a sample for amplification. Explaining each step in a quick and even tone that suggested many years of constant repetition, along the way he noted methods of conserving materials and reducing plastic waste, a persistent concern among eco- and cost-conscious biohackers.

Rockefeller lifted his three specimens up to the camera for all to see, picked on recent collection trips in California and Ecuador. Each would be put through the process of PCR amplification, the results sent off for sequencing by a nearby genomics company called Genewiz. The bar codes that came back would help to confirm the identity of the mushroom, which would then be logged in GenBank, an online repository of all publicly available genetic data, operated in part by the National Institutes of Health. To demonstrate that last part required one more screen.

While the samples underwent the gel electrophoresis step that would reveal how successful the gene-copying process had been, we were ushered to another corner of CCL, where Rockefeller opened up his laptop. After a few keystrokes in a Linux command line—testament to his open-source sensibility—he pulled up GenBank. Through the online interface, biologists, professionals, or amateurs working with university partners can prepare and contribute samples that live in the database for anyone to access. It offers a way to add to the global picture of fungal diversity, by not just submitting sequences but also comparing those already in the database, which skilled, committed mycophiles trawl to find connections among the genetic data, or correct errors.

As we watched, Rockefeller ran a BLAST—the basic local alignment search tool—used to find and compare sets of sequences within the database, and pulled up entries for a *Chromosera* mushroom like the one he'd just prepared for sequencing. Scrolling through the results, he quickly noted a minute but significant discrepancy between two sequences. "We've just uncovered evidence of a new species," he said somewhat nonchalantly. Turns out that in the under-resourced field of fungal research, new discoveries are rather commonplace, although formally naming a new species is much more involved; Rockefeller had thus far been involved in the naming of one species: *Psilocybe allenii.*[29]

Rockefeller travels the country giving demonstrations of fungal genetics, microscopy, and field identification. "It kind of pays for itself," he told me over the phone before the online class. In addition to teaching, he also spent an astonishing amount of time scouring and updating other people's records online, identifying mushrooms and adding his own, while also managing various social media pages. A fixture of the online mushroom identification platforms and communities, he credits his prominence partly to having a recognizable last name. "A lot of people think I'm some sort of rock star," he said, "but I don't think of myself that way. I'm just somebody that takes a lot of pictures of mushrooms and puts them up on the internet; I never expected to get famous for doing mushroom stuff, but it's certainly moving in that direction." Shortly after our conversation, an online petition was floated to Vice Media to hire Rockefeller as host of an as-yet-nonexistent video series about mushrooms, quickly earning over a thousand signatures.[30]

This atypical professional path into the sciences came by way of a career in cybersecurity and an interest in cracking computer systems. Lately, Rockefeller satisfies his subversive impulse by collecting specimens from distant or restricted areas of the woods, and by showing people around the country how to participate in mycology. "I definitely enjoy breaking stupid rules that shouldn't be there in the first place," he said. "If you can do something good for the world like help in a conservation effort, while breaking some laws at the same time, that's like a win-win."

Teaching people how to get genetic information from mushrooms might not sound like the most rebellious activity. But the idea that with a few bucks and the patience to learn new skills, anyone can participate in an area usually reserved for people with advanced degrees or sinecures at elite institutions does seem a little bit revolutionary. That is exactly how Rockefeller described the availability of new technologies and online platforms, and the impact they're having on citizen science.

It's easy to think of staring at a screen as the opposite of paying attention to the world around us. Technology and social media have certainly done plenty to corrode our minds and social bonds, but even the platforms so easily associated with wasted time and toxic discourse can become helpful havens for the scientifically curious.

"For all of the terrible things that Facebook does, it is an extraordinary place to be a naturalist," said Christian Schwarz, a mycologist, citizen science advocate, and self-described "biodiversiphile" based in Santa Cruz, California. "If you set up your newsfeed right, if you subscribe to the right people and follow the right things, you can actually get a social media experience that is what it promised it would be, which is a way of communicating with like-minded people about things that you care about."

Schwarz, a lanky guy with a gift for locution and a seemingly burdensome depth of intellect, studied biology at the University of California, Santa Cruz, where he discovered a new species of fungi that he named after his professor.[31] By the age of twenty-nine, he'd cowritten a comprehensive mushroom field guide for the Redwood Coast, and has spent his time since traveling the country, educating and encouraging citizen scientists to engage with nature. Fungi are far from his only focus, but they've proved helpful in attracting people to notice and document local biodiversity.

"Mushrooms are this perfect organism for cultivating engagement," he told me over the phone one afternoon. "It seems like there's a weird confluence of historical happenstance, where citizen science in and of itself, regardless of whether or not it has to do with mushrooms, is really hitting its stride right at the same time that there's this renaissance, this awakening about mushrooms in North America, especially in the West and East Coasts. And the fact that they're happening concurrently means that mushroomers are doing some of the most interesting stuff with citizen science."

Schwarz advocates for putting tools into the hands of anyone with even minor interest in the natural sciences, in hopes that might entice them to engage further. Participation can be as simple as sharing what they've found while walking around their neighborhood, as casually as one might post a picture of a cute dog or an especially attractive plate of food.

To this end, Schwarz favors an app called iNaturalist. The basic idea is simple: Anytime you encounter something interesting in the wild— plant, animal, fungi, anything alive—just pull out your smartphone, snap a picture, and upload it through the app. Along with the picture goes the location and the time of the find, viewable on a global "heat map" so that others can search by organism or region, and the identity of the organism can be deduced by the crowd. In this way, anyone can create or add to region- or organism-specific projects. Schwarz managed dozens of such projects on the platform, including collections on both coasts of the United States. Anyone who spots a mushroom near Santa Cruz, California, for example, could easily add to the more than thirty-one thousand observations in a collection called *The Mycoflora of Santa Cruz County*.[32]

Off-screen, Schwarz regularly takes budding naturalists of various stripes on walks in the woods, teaching them about the ecologies around them, encouraging them to take their own pictures and upload them to the app. But even when not on the trail, like Rockefeller, he is frequently online, on the lookout for opportunities to engage anyone whose activity suggests an interest in the natural world.

"I will often see someone post something on Facebook, because that's the platform they're used to," Schwarz said. "They're not a naturalist per se, they don't see themselves as highly attuned to the natural world or particularly interested in it, but they see something on the street while they're walking their dog, and they take a picture of it and they just upload it to Facebook very rapidly. It's not a good picture, and they just say, 'What is this thing?' And I'll say, 'Oh, wow, that thing is really cool, and here's *why* it's cool. Please consider uploading your picture to iNaturalist.' When I started doing this, it was sort of flippant, but I actually now have a hot key on my cell phone so I can just press three letters and it fills out this little message that says, 'Please consider adding these photos to iNaturalist.' I solicit this engagement, this extra step from people so often, ten, fifteen times a day, and it's amazing how often they will follow through. For some fraction of those people—it's a numbers game—they develop an internal interest that keeps going, and they start paying attention."

The key advantage offered by new, accessible, and broadly distributed technologies for nature documentation and engagement might have more to do with the scale of participation than the quality of the information. Smaller communities of highly trained mycologists make for much more accurate data, but less of it. The alternative is to get as many people involved as possible, and for a marginalized field at the beginning of a surge in popular interest, that seems an ideal approach. "Set the gates wide and the door narrow," said Schwarz of his own philosophy on the matter. "Bring in all the data and then sift it out later. You don't want to exclude it out in front."

Over time, so the idea went, recruiting an army of casual observers would make it possible to crowd-source data that's useful to the scientific community at large. The question was: how to involve a large-enough crowd, a challenge that seems to come down into creating a culture of naturalistic curiosity on the one hand and accessible tools for documentation on the other. Platforms like iNaturalist leaned heavily on the former, while others like Mushroom Observer were tooled for more rigorous data collection.

Rockefeller, for his part, spends most of his community-engagement energy on Facebook and Mushroom Observer, which includes tools for generating consensus about identifications, and integration with existing herbaria and fungaria as well as other databases like MyCoPortal and MycoBank, tracking the ongoing changes in species names and other features probably most useful to those with more robust knowledge and experience than those just getting started. "Mushroom Observer and iNaturalist are definitely marketed more toward amateurs, but professionals will use them sometimes," said Rockefeller. "I'll go into iNaturalist and sit down, and in two hours I'll correct five hundred records. And, you know, that makes a lot of people very happy, to know what they found. If there were a thousand people doing the same thing, it could be kept up with, but there's only like a handful of us that actually do much identification on there."

Hundreds of thousands of people share daily mushroom pictures on the range of social media platforms, which probably does more to build community and interest around the subject than to generate a useful data set on fungal diversity and distribution. But nevertheless, it represents a mass engagement, however casual, which sometimes leads to meaningful discoveries. In 2020, for example, pictures shared on social media led to the identification of a new species of fungus, named aptly, *Troglomyces twitteri*.[33] Perhaps just as important as gathering scientifically valuable data, though, is fostering a culture of participation and curiosity about nature that's inviting to anyone, no matter their level of expertise.

The idea for iNaturalist grew out of a graduate project by Nathan Agrin, Jessica Kline, and Ken-ichi Ueda at the UC Berkeley School of Information, initially with an emphasis on birders and mushroom hunters.[34] It launched in 2008 and grew fast, with 42 million observations (2.2 million of them fungi) and 218,000 observers as of June 2020. The nonprofit produced a video to celebrate its 25 millionth observation, showing the spread of its use over twelve years, depicting a gray map of the world flickering with red dots that grew into larger and larger clusters, until they blanketed the whole screen.[35] Considering the global pandemic gaining steam at the very moment I first watched it, the imagery was a tad unsettling, but actually represented something quite positive: a diverse community quickly forming across borders on every continent, composed of people connecting over common interest in the natural world.

To those running the platform, at least as important as the quality of the data generated by all those people is the participation itself. "We often say 'observe locally, identify globally,'" Tony Iwane, outreach and community coordinator at iNaturalist, told me. "Our main goal is to help people develop a habit of nature observation and documentation whenever you're outside, to think about what's around you, to notice what's around you, and be curious about it and want to share it with people."

Generating scientifically usable data is something of a secondary goal of the platform, which is powered by its social networking dimension. Users naturally cluster around their local regions, or favorite organismic groups, submitting, classifying, and refining one another's observations. This on its own could be enough to raise awareness; in the best-case scenarios, these communities and the records they generate can serve conservation efforts, arguably the top benefit of documenting biodiversity. "When people start noticing these things, they'll want to protect these things," said Iwane, noting a student marine conservation group in Mumbai that had, at least temporarily, halted construction of a coastal highway with the biodiversity information collected on the platform.[36]

With enough people participating, and enough expertise among them, it may not be long before they can reliably generate valuable information for all manner of scientific projects. But that requires a culture of widespread, sustained, and productive engagement with the natural world. Perhaps paradoxically, this, too, is something social media can help accomplish. "A lot of people call iNaturalist 'citizen science' and we sort of push back against that," said Iwane, adding that the company, now backed by a partnership with National Geographic, sees its platform's primary role as being that of bringing the bottom-up dynamics of social network to the often top-down arrangements of citizen science. If they perceive an opportunity to serve as a scientific resource, it is by bringing together folks who might not otherwise have the means of collaborating, nor the inclination.

"I think a lot of people that use iNaturalist are maybe a little more introverted; a lot of us want to be around a fungus or a lizard more than we want to be around other people. But we still do want to be around other people."

———————

When the Fungal Diversity Survey—FunDiS, formerly the North American Mycoflora Project—officially launched in 2018, its goal was to activate amateurs across North America in building out a map of the continent's mushrooms. "Knowing what exists, where it exists, and when it exists is vitally important," said Stephen Russell, a mycologist working toward his PhD in botany and plant pathology at Purdue University. When I spoke with him, he was secretary of the FunDiS and one of its lead organizers; we met while he and his partner were running a collection table at a mushroom festival in rural Oregon. Forayers returning from the woods were being encouraged to document them and submit samples of tissue for sequencing, all as part of the effort of documenting the region's macrofungi.[37]

"We have a lot of data on plants, for example, and so scientists have looked at tree migration over time. But there's very little with fungi, and that's primarily because we don't have an understanding of what even exists out there at a base level."

FunDiS encouraged anyone going on a foray to photograph what they found, and to preserve and prepare specimens for sequencing wherever possible. The project was built out with the use of iNaturalist and GenBank, MycoPortal, and a prototype documentation site called MycoMap, so that the records could be easily created and accessed by almost anyone. "Basically anytime that you go out on a mushroom hunt, there is a very good likelihood that you're going to be finding multiple specimens that are currently undescribed in the scientific literature," said Russell. "It all starts with that observation of a mushroom out in nature."

To connect with collectors, FunDiS coordinated with the North American Mycological Association, the country's largest amateur mycological organization, host to some eighty chapters throughout the United States and Canada. They also partnered with the Mycological Society of America, the professional association for mycologists. At various mushroom festivals and other events, they ran tables where mushroom pickers were invited to submit their samples, which were in turn sent to Purdue University for sequencing. Amateur mycophiles were also encouraged to set up their own projects for documenting, sequencing, and vouchering the fungi in their region—to date, there are 165 of these projects, reaching from Alaska to Puerto Rico, Newfoundland to Hawaii.[38] For the older set that tends to define the populations of the mushroom clubs, vouchering and uploading images through an iPhone app may not be entirely intuitive; for a younger, online generation of mycophiles, it's second nature.

In some ways, the culture of the professional mycologist and the mycologically curious is becoming more distinct. Specialization is a trend in all sciences, and mycologists tend to focus on a single species or group of species, using highly technical and often expensive methods in genomics or data analysis to mine for ever more granular insights. Gathering mushrooms in the field is not a purely scientific activity, but as Kew's LAFF project demonstrated it can be invaluable to science. People of all ages and persuasions are already doing the collecting, and their numbers are growing; if those efforts could be tuned to provide the basic information of fungal diversity and distribution—which mushrooms are where, and when—it could help to establish a mutually beneficial (some might say *symbiotic*) relationship between those who wear lab coats and those who wander into the woods (assuming they're not the same person). This is the dream of citizen science in any field, but accomplishing it often takes money and organizational resources. Fostering interest in fungi that organically scales into a sustainable resource to science and citizen alike is the challenge facing the FunDiS.

"The kind of citizen science project that we are might be described as bottom-up, while the vast majority are top-down," said Bill Sheehan, president of the FunDiS. In the more common top-down example, he said, "Basically you start with money and experts, usually in academia, and they figure they want to get data collectors. And so they design and run the project, and citizen scientists go out and monitor the air or the water or any of a hundred other things."[39]

What FunDiS hopes to do, by contrast, is develop and promote a simple, accessible framework that anyone already interested in fungi could engage in order to make contributions that would benefit science, driven primarily by the interest of individuals and their communities. "Getting a good document, voucher, and sequence is kind of the ultimate trinity," said Sheehan, "although I see different stages of citizen science in our project."

For FunDiS, Sheehan drew up a four-stage plan, basically a mycological ladder of engagement.[40] Stage one involves reaching as many people as possible on social media, inviting them to record fungi in the field and upload them to platforms like iNaturalist. With many thousands of people interested in mushrooms and fungi on all kinds of platforms—several Facebook pages have tens of thousands of active members; Sheehan cites a single group with some 180,000 members, growing by the thousands each month—there is a ready audience of people already sharing and engaging. The trick would simply be to funnel that interest into scientifically valuable activities.

Stage two involves inviting participants to prepare and submit mushroom tissues for DNA sequencing, using techniques like those Rockefeller was demonstrating. (Sheehan described him as "the Johnny Appleseed of DIY sequencing.") In stage three, dried specimens are saved in a fungarium, as Schwarz has managed to do thanks to a special arrangement with UC Santa Cruz, although some mycologists also operate their own at home. The fourth and final step addresses what Sheehan calls "super users," like Rockefeller and Schwarz, to teach others the more advanced concepts and methods, and to engage and refine the data submitted by the first three tiers of participants.

"There might be a hundred people in academia who are doing this stuff," said Sheehan. "Say one hundred thousand people are out recording during stage one, and maybe ten percent of those are curious enough to want to find out what the DNA is for their specimens, so you get ten thousand to submit tissue. If we had maybe ten percent of that ten thousand doing the vouchering as well as the DNA submission in the field recording, now you're down to a thousand, maybe, and then the stage four is you engage and train the super users in analyzing the DNA results and maybe also doing part of the DNA sequencing at home or in class."

Sheehan bemoaned financial difficulties as the organization struggled to find a solvent model. Funding from national grants was hard to come by, and seeking the support of, say, a pharmaceutical company would risk flipping their model into the sort of top-down arrangement that FunDiS sought to avoid. The approach Sheehan wanted to advance involved creating a community; as he put it, to emphasize the *citizen* as much as the science.

Mycology is rather unique among the natural sciences for its sizable base of dedicated, active amateurs throughout the country, who regularly go out and document fungi as a part of daily life. Birders are perhaps the closest analogue, and represent a massive community indeed. One advantage of mushrooms, though, is that they don't require luck or quick reflexes to take good pictures of them, or to collect them. Also, the presumed diversity of fungal species dwarfs that of birds; estimates put the total number of bird species at ten thousand to eighteen thousand.[41] That's fewer than the entirety of named mushrooms, only 144,000 of which have been named.[42] That creates a huge gap of knowledge to address in the latter case. But birders also have a technological advantage: the app eBird is host to some 737 million total observations, more than 140 million posted in 2019 alone, all just birds, voluminous data that has informed more than three hundred peer-reviewed papers.[43] Contrast that with just over two million fungal observations on iNaturalist—for an entire kingdom versus a single class of animal—and the gap is striking. But so is the opportunity to unify and instrumentalize the efforts of a vast, diverse community of people passionate about fungi.

In August of 2019, FunDiS sponsored a Continental MycoBlitz, an organized effort to collect two thousand specimens throughout the country. For one week in the summer and another in the fall, the platform made a nationwide push to expand its engagement, at festivals and forays across the country, but mostly by individuals, leaning heavily on social media to get the word out. By the end, the event's iNaturalist page listed 5,301 observations comprising 1,169 species, predominantly along the coasts and upper Midwest.[44] To Sheehan, it represents a promising model for documenting the vast fungal diversity throughout North America. "I see it as a great untapped opportunity for citizen science," he said, "because Fungi are the neglected kingdom."

CHAPTER FIVE

THis LanD is MYcoLanD

Approaching Colorado's San Juan Mountains from the desert
to the south, one sees chaparral-dusted mesas and neon stripes
of sedimentary layer cake give way to rolling hills that rise into
aspen forest. Over the course of a few hours, you approach ten thousand
feet, passing pastures and canyons and the occasional whiff of sulfur
springs. Eventually, a hairpin turn in the corner of a steep valley reveals the
face of Wilson Peak, the majestic crag emblazoned on cans of Coors beer.

At dusk, I made it to Telluride. Nestled at the base of a steep canyon,
the mining-town-turned-ski-resort plays host each summer to one of
North America's best-known mycological happenings, the Telluride
Mushroom Festival. The (officially) five-day-long gathering is a chance
for the mycologically inclined to partake in forays, lectures, panels, and
workshops held in a handful of venues scattered around picture-perfect
Colorado Avenue. During TMF, the main drag is bisected by a huge
banner announcing the event and a long strip of vendor tents. Attendees
and passersby alike can swoop up shirts, sculptures, tarot cards, books,
magazines, jerky, foraging knives, more mushroom swag than you could
shake a cane at. (Mushroom canes are also available.) The roughly six
hundred people who bought tickets in 2019 attended hands-on cooking classes, cultivation workshops, screenings, outdoor tours, an "open
myc" night, and the popular MycoLicious MycoLuscious MycoLogical
Poetry Show.[1]

Every morning, organized forays set out to spend the daylight hours plucking mushrooms from the ample hills and valleys that surround the town. Slots are limited and fill up quickly, so hopeful fungus hunters were advised to arrive up to two hours early, gathering in the dawn light before departing in improvised caravans, returning in the evening with full bags and wide grins. Any mushrooms not bound for the skillet were left at an identification station in tiny Elks Park. Under the shade of a medieval-style pavilion tent, experienced pickers mused over the specimens, hands placed pensively on their chins, while kids peeked over the table edges to cautiously poke at the various piles. When dropping off a mushroom, one was expected to write down distinguishing factors on a small slip of paper, such as where it was found, so that a volunteer team of identifiers could properly sort them. Occasional grumbles suggested many never bothered to fill out the cards, perhaps a sign of all the newbies in attendance. (Mea culpa: I simply left all the mushrooms I'd found in a heap on the corner of an ID table, only later realizing my foray faux pas.)

Just outside the tent, people danced around a thrumming drum circle in a scene as evocative of a rave as of a Grateful Dead concert. Revelers were costumed in fungal fashion. Some wore toadstool hats reminiscent of the character Toad from *Super Mario Bros.*; a trio of visitors lurked in full-body veils in eerie imitations of the dead-man's-fingers mushroom. Leif Olson, an environmental scientist and educator, swayed to the music wearing shredded white cloth, his improvised ode to a lion's mane mushroom; Giuliana Furci grinned from behind circular sunglasses and underneath a floral headdress sprouting a cluster of knitted ink cap mushrooms. Both would later be presenting lectures and workshops at the festival.

In some ways, TMF might easily compare to Burning Man, and the event is in fact timed so as not to leave people facing a choice between the two. The two festivals pose similar challenges: a ticket to TMF sells for hundreds of dollars, and the costs of travel and accommodations can easily run into the thousands. Those who can afford to attend, however, seem to agree it is money well spent: a chance to learn, foray in a gorgeous landscape, and commune with their fellow mycophiles.

To the outside world, the festival is most recognizable for the parade that takes place on its penultimate day. For several surreal hours, a cavalcade filled the streets with dancing, drums, fantastic costumes, and, bobbing above the crowd, shameless mushroom puns on homemade signs, some with political overtones. (DOSE TRUMP read one; NON-JUDGMENT DAY IS COMING read another. My favorite depicted crudely drawn UFOs dusting the Earth with spores. The picture was a reference to panspermia, the surprisingly popular theory which says that life on our planet was seeded by spores from outer space. Next to it was a playful take on the acronym for NASA: NEVER A STRAIGHT ANSWER.[2]) Marching at the head of the throng was an impish, elderly fellow with a long beard, top hat, cape, and staff, all festooned with mushroom iconography: Art Goodtimes, the Telluride Mushroom Festival's poet-in-residence and hippie hype man.

Press pictures of the procession invariably feature Goodtimes and his road-worn, red-and-white speckled Toyota pickup truck, its bed weighted down with a giant re-creation of an *Amanita muscaria* painted in the same colors. My first interaction with Goodtimes occurred when he recruited me, along with a handful of onlookers standing on the sidewalk, to help push the truck to its starting point. "Hopefully the brakes work!" he laughed. To my great relief, the car stopped just fine, and as the parade geared up, the picturesque corridor of Colorado Avenue took on a delirious energy in the bright August afternoon. Goodtimes ambled around the parade's perimeter. Stopping at the edge, his arms in the air, he led the crowd in a loud countdown of: "We! Love! Mushroooms!"

———

The Telluride Mushroom Festival celebrated its thirty-ninth year in 2019, a run in which it grew from an obscure conference focused on all aspects of mushrooms, particularly mind-altering fungi and compounds—commonly grouped together as "entheogens"—to a high-profile, fungal bacchanal. It is the very picture of fringe mushroom culture, a psychedelic trip unleashed on Main Street. In 2019, TMF was even the subject of a photo essay in *Vogue*.[3] Like all life, though, the culture around fungi is evolving, and in directions that aren't quite so on the nose as a parade of mushroom costumes or a gaggle of tripping hippies.

Dashing between venues and meeting places, I scribbled notes in dark theaters as experts relayed the state of research into medicinal compounds produced by fungi; the progress of clinical trials of psilocybin for treating addiction; briefings and site tours for projects involving mycoremediation and mycoforestry—that is, working with fungi to degrade toxins in soils and encourage more fire-, drought-, and erosion-resilient forests. These heady sessions contrasted with classes on cooking, dye-making, consultations with a "psychedelic psychotherapist," and other offbeat offerings. TMF represented a certain cultural intersection of the old and the new in mycology, much of it connected by the legacy of psychedelic mushrooms.

"For the first twenty, twenty-five years or so, it was really the country's only entheogenic conference going on, but it wasn't promoted exactly like that," Goodtimes told me after the festival. We met on the back patio of a house on the edge of town, where a small crowd had gathered in celebration of his seventy-fourth birthday. "We talked about LSD, we talked about MDMA, ecstasy, we talked about ayahuasca, we talked about toad venom. So there were lots and lots of discussions about various entheogens for those first twenty-five years, and it was kind of under the radar."

The term *entheogen* was coined in the mid-twentieth century as a less loaded alternative to *hallucinogen*, the former translatable to "the divine within."[4] As popular interest in entheogens and fungi has grown, so has the festival's resonance with the mainstream.[5] TMF's rising profile came in part by way of the legitimization of research into therapeutic applications of psilocybin-derived compounds, particularly for end-of-life care. Institutions as prominent as Johns Hopkins began conducting clinical trials on psychedelic therapies in 1999, opening a research center dedicated to psychedelic (literally, "mind-manifesting") therapies in 2019.[6]

Being under the radar didn't mean being off it, though. "We had DEA agents come when Sasha Shulgin was here, I know for sure," said Goodtimes.[7] "There were DEA agents along the way, but nobody ever gave us any trouble. I was always surprised that we didn't actually get more heat, having come from the Bay Area and seeing the infiltration of various groups from the Black Panthers to Earth First!, which I was also involved in."

Surrounded by mushroom paraphernalia, dreadlocks, drum circles, tie-dyed saris, and folks of a certain age singing to a keyboard performance of "All You Need Is Love," a person might easily have read the scene as an echo of the bygone era of free love. But the intention at TMF is clearly to tap into something deeper than "shroomer" culture.[8] The conversations, activities, pageantry, eccentricity—all are part of what Goodtimes characterized as a collective ritual. "When we're mushroom hunting and going on forays, we're participating in a very ancient thing," he told me.

Naturally, the ritual aspect encompasses the consumption and use of mind-expanding substances as well. Traditions and practices around entheogens trace back centuries and extend throughout the world. Through shamanism, song, dance, and untold other practices, various cultures have developed productive ways of engaging and making meaning of the challenging and reorienting experiences these substances—commonly characterized as medicines—inspire.[9] Despite widespread Indigenous use of entheogens throughout the Americas, these practices were largely unknown to European-descended North Americans until the mid-twentieth century, leaving a short time in which to develop our own rituals and practices around their use. Terms like *set and setting* have since become immutable truths among those who take the plunge into psychedelics, referring to whom you trip with, where, and in what frame of mind. It's hardly ancient wisdom or the basis of a robust system of practice, but it's a start, and many psychonauts have well-developed philosophies that extend far beyond the set and setting maxim. Part of the hope of a gathering like TMF was to improve and refine our collective relationship to the psychedelic.

"In Indigenous use of entheogens, they always had a matrix, it's always embedded in a cultural system of some kind," said Goodtimes. "Unfortunately, we don't have that, and so people either take it in a laboratory setting, or they take it on the street, and neither one of those are exactly the optimum kind of conditions. So I think we're still struggling to figure out, how can we incorporate it into the culture in a sensible and responsible way?"

Ethnomycologists study the entanglement of human culture with fungi, including the psychedelic variety. The often hypothetical history of mushroom use articulated by the early ethnomycologists inspires and informs much of the modern North American perspective on psychedelic best practices, in a context where substances like cannabis, salvia, and of course psilocybin mushrooms are scheduled as illicit substances, with heavy penalties levied for their use and especially their distribution (although several states have in recent years taken significant steps toward decriminalization).[10] The "magic mushrooms" now synonymous with psychedelic spiritual practice and counterculture were introduced to North America only in the mid-twentieth century, along with LSD. That's meant less than a century for our modern culture to develop concepts and traditions around psychoactive fungi, which societies older than postcolonial North America—as most are—have had millennia to cultivate.[11]

To tell someone that you're into mushrooms is to invite a raised eyebrow and perhaps some cautious questions about your relationship to the law: "So are you into *magic* mushrooms?" Many mycophiles have learned to keep ready the reply, "*All* mushrooms are magic!" But before May 13, 1957, fungi and illicit activity weren't so intuitively linked in the average American's mind. That's when the nation was introduced to the existence of so-called magic mushrooms by way of an article in *Life* magazine penned by one R. Gordon Wasson, former vice president of public relations for J.P. Morgan & Company. The article, titled "Seeking the Magic Mushroom," recounted his 1955 visits to Huautla de Jiménez, a small village in the southern Mexican state of Oaxaca. There, Wasson and photographer Allan Richardson became the first outsiders known to undertake the sacred Velada ritual, under the guidance of the Mazatec curandera Maria Sabina, and her daughter, Maria Apolonia. Making several visits to the remote village over the course of four years, Wasson convinced a reluctant Sabina to include him in the sacred spiritual practice under the pretense of concerns over his son's safety.[12] He had journeyed to Huautla following rumors of sacred mushroom rituals still practiced there, part of a broader search for the fungal foundations of spiritual practices the world over.

"Was it not probable that, long ago, long before the beginnings of written history, our ancestors had worshiped a divine mushroom?" wrote

Wasson at the outset of his fourteen-page article. "This would explain the aura of the supernatural in which all fungi seem to be bathed. We were the first to offer the conjecture of a divine mushroom in the remote cultural background of the European peoples, and the conjecture at once posed a further problem: what kind of mushroom was once worshiped and why."[13]

In the years leading up to Huautla, Wasson and his wife, pediatrician Valentina Pavlovna Guercken, rather unexpectedly became committed and influential ethnomycologists. Their peculiar passion was ignited, the story goes, when one night Valentina, who was from Russia, returned home in an excited state after discovering a trove of edible mushrooms on a forest walk. She announced her plans to make dinner from them, to the horror of her husband. "Not long married," he recounted, "I thought to wake up the next morning a widower."[14] Culinary preferences aside, their reactions suggested deeply set and widely divergent opinions of mushrooms, a difference they found profoundly intriguing. "What pre-historic religion or tabu is finding expression when the English governess, with a facial spasm and shudder, grinds a delectable mushroom under her heel and warns away her charges?" Wasson wrote in a note to his colleague Robert Graves.[15] The urge to kick mushrooms, it would seem, is nothing new.

The couple soon developed the belief that most cultures can be broadly defined as "mycophilic" or "mycophobic."[16] To the extent that mycophobia is a real and discrete phenomenon, its origins are difficult to pin down. In North America, it may be partly a European inheritance. In the U.K., lore abounds about "fairy rings," the suspiciously perfect circles of mushrooms that expand in grassy fields. Like a sort of miniature, medieval version of crop circles, they were thought to trace the paths of fairies as they danced overnight—stepping into the middle meant entering the realm of the devil. What once seemed a bad omen is in fact the sign of healthy mycelium growing in a radial pattern.[17] Shakespeare famously invoked them in the final act of *The Tempest*, as Prospero addresses the fairies in renouncing his diabolical magic:

> *You demi-puppets that*
> *By moonshine do the green sour ringlets make*

Whereof the ewe not bites; and you whose pastime
Is to make midnight mushrooms, that rejoice
To hear the solemn curfew[18]

Wasson and Guercken's studies would lead them to focus on South and Central America, where sacred mushroom practices were still extant despite attempts at repression, though the mushrooms themselves may have been less a factor than the language and practices around them. The name given to the mushrooms—*teonanácatl* (or "god's flesh")—was not viewed favorably by Spanish colonizers with their Catholic sensibilities. Franciscan friar Toribio de Benavente observed the Aztec practices personally, relaying a horrified tale of hallucinating heathens begging for death.[19] Afterward, a manual for the clergy was published with the recommendation that priests taking confession from Indigenous people include a question about whether they'd eaten or even served the offending mushrooms, later named *Psilocybe mexicana*.[20] The mushroom stones of Guatemala—carved humanoid figurines with large mushroom caps atop their heads—suggest fungi-related spiritual traditions may have resonated well south of Mexico and into Mayan territory.[21]

The emerging history uncovered by early ethnomycologists—among whom the Wassons became most prominent—was tantalizing, but entheogens weren't exactly new to science in 1957. Albert Hofmann first isolated lysergic acid diethylamide (LSD) nearly two decades earlier, in 1938, while prospecting for circulation-enhancing molecules to aid in the Allied war effort. In a story that has since become legend, the scientist accidentally dosed himself at Sandoz Labs in Switzerland, then somehow managed to make it home safely. There, he underwent the first documented LSD trip, albeit unintentionally, free of expectation or preconception.[22] Understandably intrigued, he began conducting rigorous research, including into mushrooms—LSD itself was derived from an ergot fungus—and ultimately identified their active ingredients: psilocybin and psilocin. LSD became well known and even popular among a community of in-the-know scientists, their subjects, and associates. Within that circle was Henry Luce, founder and editor-in-chief of Time-Life. The well-connected magazine tycoon and his wife, Clare

Booth Luce, had begun enthusiastically experimenting with LSD in the 1950s. So when Wasson approached *Life* magazine with his story from the remote reaches of Mexico, it was received warmly.[23]

At the time, the United States was not yet in the throes of "tune in, drop out" countercultural upheaval; this was the year that *The Ed Sullivan Show* refused to show Elvis Presley below the waist. The magazine that landed at the doorsteps of more than five million Americans had on its cover the familiar face of Bert Lahr, best known as the Cowardly Lion in *The Wizard of Oz*, peeking mischievously from behind some palm fronds. But despite the wholesome image on its cover, the contents were wholly unfamiliar to most and, intentionally or not, downright subversive. Wasson's feature in the popular magazine came complete with what amounted to a trip report, along with photographs and illustrations of various species of psychoactive mushrooms.[24] It was essentially a field guide for what would soon come to be classified as a Schedule I narcotic. In short order, Oaxaca—and particularly Huautla, high up in its northern mountains—became a hotbed of hippies and trip hunters.

The '60s saw LSD and mushrooms solidly associated with an ascendant counterculture. For many, true initiation meant making a pilgrimage to Mexico, to the extent that in July 1969, the Mexican government cracked down, forcibly evicting nearly one hundred of the "vicious hippies" on charges of drug trafficking. The local paper *El Universal* warned of "contamination of our youth" by North American *jipis*.[25] Maria Sabina herself expressed regret for having shared the ritual with Wasson, over what she perceived as the diminishment of the mushrooms' magic. By 1971, the United States had declared that psilocybin represented as much of a threat to society as heroin or crack cocaine, and mushroom subculture largely went underground.[26]

Productive Strains

The United States is widely regarded as the epicenter of the hippie movement of the 1960s, during which mushrooms gained much of their modern cultural cachet, for better or worse. But the first definitive, recorded recreational use of *Psilocybe* mushroom growing outside of

Mexico actually took place in Canada, when Royal Canadian Mounted Police in Vancouver seized liberty cap mushrooms from a group of college students.[27] "We sort of presume that somebody who was going to University of British Columbia went down to Mexico and experienced them there," said mycologist Paul Kroeger. "Then, when they came back here, noticed mushrooms that looked very much the same, and tried them."

I met Kroeger at an evening meeting of the Vancouver Mycological Society. He is a charter member, since its founding in 1979, a world-class mushroom identifier and author of several studies documenting the region's fungi. Prior to the founding of the VMS, "Vancouver was sort of in the forefront of all the stuff, and the shenanigans that were going on," said Kroeger in a soft, rumbling voice inflected with the classic Canadian raise, his long gray beard and ponytail alluding to his own relationship to the scene he described. "We had our own little district, our version of Haight-Ashbury or Greenwich Village, which was the Fourth Avenue."[28]

Liberty cap mushrooms, a *Psilocybe* variety common to the U.K., were the first to gain popularity in British Columbia as psychedelics entered public consciousness. One theory suggests that the fungus caught a ride with livestock or fodder grasses from Europe.[29] It's a truism that mushrooms, and particularly *Psilocybe* for some reason, travel well and grow abundantly on the edge of human-disrupted environments. As a wave of redevelopment began to reshape Vancouver and farther along the Pacific Northwest, they were known to pop up, a bit ironically, near newly built police stations.[30]

Amanita muscaria is also common to British Columbia and the Haida Gwaii islands to the west. But there is another amanita, one that immigrated to BC, which can offer a one-way trip for those who eat it: *Amanita phalloides*. Known best as the death cap, these mushrooms wreak havoc on the liver of anyone unlucky or inexperienced enough to consume one. *Phalloides* are believed to have come in on trees imported from Europe, gradually hitchhiking up the West Coast from California, where they were first detected in the 1930s.[31] *Phalloides* might be considered an "invasive fungus," but after all, it seems to travel largely as a result of human activities. Since their discovery under sweet chestnut trees near Vancouver in 1997, they've poisoned six people, one of them a young child who, tragically, did not survive the encounter.[32]

Much of the fear and disdain that exists toward fungi can naturally be attributed to their poisonous potential, even though the vast majority of mushrooms are not deadly; there are many more that taste like heaven than those that will send you there. Nevertheless, only in the company of an experienced identifier should you even entertain the notion of eating a wildcrafted mushroom.

For the sake of public safety, in 2014 Kroeger and the VMS set about systematically documenting the spread of the death cap. Being a mycorrhizal fungus, it grows only in the presence of certain trees, and was first known to pop up near the hornbeams planted en masse throughout the area during a wave of development. Documenting death cap distribution entailed scouring records to see where hornbeams had been planted, and sending VMS members to visit those locations to note which ones hosted the hazardous mushrooms. Members also kept an eye out during their daily travels around town, always on alert for the slender white stalks emerging from the egg-shaped volva common to amanitas, with their distinct pearlescent green caps. Once you've learned to identify it, *Phalloides* carries the air of a tainted apple from fairy tales.

Eventually, the mushrooms acclimated to their new environment, and the list of associated trees grew longer. By the time we spoke, Kroeger said, the mushroom had jumped to the Garry oak, a tree native to the British Columbia coast, and was deemed invasive to the area.[33] That made the systematic, tree-specific surveys pretty much moot. "We've got gobs of records now," he told me. "We just record them as they're found." Every mushroom season, in coordination with public health authorities, the VMS issues service notices reminding people that mushroom picking can be dangerous.

Kroeger, who made his living identifying mushrooms, as a kind of fungal consultant for scientific research projects, found more and more of his time spent assisting poison control agencies. "We used to have a lot of inquiries, people calling up and saying, 'Oh, I have these mushrooms growing in my yard, can I eat them?'" he said with a chuckle. "Now we get many more calls saying, 'We've got these in our yard, are they dangerous?'"

As an example of why so many are wary of fungi, the image of deadly mushrooms popping up throughout an entire city is a compelling one. But even while public concerns over poisonous varieties have risen, popular interest in them has steadily grown as well. Kroeger noted this as VMS membership levels held steady while those of other naturalist and garden clubs shrank. "We've noticed that when the economy gets a bit rough, there are still lots of people interested in the mushrooms, but people don't tend to be as interested in their begonias and their roses, et cetera. And then of course a lot of garden clubs sort of presume that people have gardens, and now more and more people are living in apartments and condos, and don't have gardens, but people still like to go out and look at mushrooms."

Mycological societies began proliferating in North America in the 1960s and '70s, structured in a manner similar to their British forebears, emphasizing finding mushrooms in the woods, identifying them, and eating the tasty ones.[34] The first in the United States popped up back in the late nineteenth century; clubs started gaining prominence after the poisoning death of an Italian diplomat in Washington, DC, incidentally, after eating an *Amanita muscaria*.[35] The New York Mycological Society is perhaps the most famous, founded by avant-garde composer John Cage, emerging in its current form in 1962, out of a class he was teaching at The New School in Manhattan.[36] Gary Lincoff, the late and beloved urban forager and mycologist, was a member of the NYMS.

The VMS, too, is a well-known North American club, in part because of the incredibly rich ecological zone in which it operates. From coast to coast, there is a common culture and set of practices among these clubs, but its character is evolving as a new generation joins the often-aging ranks of clubs that gained popularity in the time of psychedelic counter-culture. Anecdotally, a flush of young members seems to be rising with a wave of interest in the potential of fungi to provide food and medicine, heal landscapes, and otherwise "save the world."

"There's much more interest in other aspects of mycology that didn't used to be the primary focus," said Kroeger, who relinquished his decades-long presidency at the VMS, refusing to return to the role so that younger members could step up to take the reins. "They used to be primarily about picking stuff to eat, but now there's much more interest in cultivation, using them for dyes, medicine, all those other aspects; mycorestoration, all that kind of stuff. So it's just as well that we have younger people coming in and sort of adapting the club to the current interests." The newest president of the VMS, Sam Reeves, had to miss the meeting I attended, as she was taking part in ongoing demonstrations of solidarity with the people of the Wet'suwet'en Nation, whose land had been seized by RCMP to make way for an oil pipeline.[37] In her stead, another young person, Mendel Skulski, took over the duty of officiating the meeting. Kroeger hung out near the back, answering questions about the various mushrooms on display and projected onto the wall.

After the meeting, Kroeger was showing off a jacket given to him by Paul Stamets, someone who has perhaps done more than anyone to influence today's popular interest in mycology. The famous myco-evangelist lives and works in the Pacific Northwest, but is known far and wide for his books and lectures as well as for his company, Fungi Perfecti. Easily the most recognized mycologist on earth, he is also an influential figure in the emerging field of applied mycology and, originally, the cultivation of psychedelic mushrooms.

Stamets's books, in particular 2004's *Mycelium Running*, 1993's *Growing Gourmet and Medicinal Mushrooms*, and 1983's *Mushroom Cultivator*, are considered standard texts among the "bemushroomed," a term he's fond of using.[38] His dazzling speeches can be readily found online, their many millions of cumulative views testament to his role inspiring an emerging generation of mycophiles and mycologists. Indeed, it was Stamets's widely viewed TED talk, called "6 Ways Mushrooms Can Save the World," that first cracked open the door for my own entry into fungal fascination.

Stamets's pitch to the public is sprawling in its scope, assessing fungi not just as interesting and useful organisms in their own right, but as representatives of the deep entanglement of all life within nested layers and levels of weblike structures. The apparent echoes of the patterns of mycelia, neurons, the internet, and the very distribution of matter throughout the cosmos, serve for Stamets as a poignant indication of the ultimate oneness of all things.[39] Fungi are often equated in his speeches with "nature's internet," and he is unambiguous in his belief that this "mycelial matrix" is quite literally sentient.

Though psychedelic mushrooms have always been fundamental to his work and life story, in recent years Stamets has begun speaking more freely and openly about the subject, thanks in part to the growing interest in psilocybin as a therapeutic, and the popularization of the subject in books such as Michael Pollan's *How to Change Your Mind*, in which he was thoroughly profiled.

A bearded, bespectacled, erstwhile-hippie-turned-eco-entrepreneur, often seen wearing a hat fashioned from the skin of a *Fomes fomentarius* mushroom, Stamets deploys a virtuosic verbosity that deftly braids the scientific and the numinous, managing this with a lifelong stutter (the overcoming of which he credits to a transformative psilocybin experience). In addition to being a highly effective communicator, he has also conducted his own enticing research into applied mycology, from cultivating fungi that can wipe out termites to breaking down motor oil with oyster mushrooms, to curing bees of colony collapse disorder with mushroom extract.[40] Though none of these mycotechnologies have yet saved the world, their example has drawn countless people into the field in hopes of realizing that goal.

Indeed Stamets's profile has risen in recent years, and he has become a recognizable figure even among those who know little or nothing about fungi. In 2019 he costarred with the mushrooms in a highly successful independent film called *Fantastic Fungi*, directed by time-lapse cinematographer Louie Schwartzberg. He teaches in-demand, pricey workshops and gives full-capacity talks at universities and fringy festivals and symposia. His influence even extends into deep space . . . well, sort of. In a recent reincarnation of the *Star Trek* television series, the titular starship *Discovery* is powered not by a warp drive, but a *spore* drive, which taps into the "universal mycelium" to zip about the galaxy.[41] Lest the provenance of the fictional technology go unnoticed, the ship's science officer is named Paul Stamets.

Stamets can seem to represent a supermassive, almost inescapable center of gravity in the public understanding of fungi. He is a prolific collector of patents, and some perceive his urgently professed mission of allyship with fungi to be overshadowed by his entrepreneurism.

There is also some reluctance to cede leadership of a growing movement for reconnection with nature to yet another well-off white dude with apparently protectionist proclivities. But even the most ardent critiques I've heard acknowledge Stamets's contribution to the popular interest in fungi, and the energizing influence he's had on the public understanding of fungi, which began with the cultivation of *Psilocybe* mushrooms.

————————

As interest in experiencing *Psilocybe* mushrooms spread in the wake of Wasson's article, supply soon became a restrictive factor. Sure, you could find them growing in the wild if you knew what to look for, but that was dependent largely on luck and the whims of climate. For intrepid psychonauts like Terence and Dennis McKenna, this arrangement would simply not suffice.

In the early '70s, the McKenna brothers paid visits to Colombia, to indulge in wide-ranging and often wild experiments with *Psilocybe* mushrooms.[42] Back in Northern California, convinced of the need to plumb the depths of noetic meaning they'd encountered in psychedelic states, they began developing accessible, reliable techniques for growing the mushrooms, kicking off a multifarious culture of cultivation and experimentation that is still alive and evolving to this day. With the development of these techniques, or "tek," pretty much anyone who wanted to could produce homegrown *Psilocybe* mushrooms, the cultivation of which was seen as much more than a technical pursuit. Indeed, it was nearly a ritual unto itself, a process by which to amplify the essential message of the fungi, to spread the "psychedelic technology" far and wide.

"I really believe in growing mushrooms," Terence said in a 1987 lecture at the California Institute for Integral Studies. "Growing the mushroom teaches you cleanliness, punctuality, attention to detail, steadiness, all of these virtues which are the very virtues you need to travel smoothly in that dimension."[43] In 1975 the McKennas wrote the first popular guide to *Psilocybe* mushroom cultivation, *Psilocybin: Magic Mushroom Grower's Guide*. It was published under the pseudonyms O.T. Oss and O.N. Oeric (a play on the words *otiose*, which means "serving no practical purpose or result," and *oneiric*, meaning "relating to dreams or dreaming").[44] The book begins with a short recitation of history and a lofty mission statement:

> *More than twenty-five years have passed since Albert Hofmann isolated and named the hallucinogen psilocybin. Hofmann's psilocybin was extracted from various species of mushrooms whose accurate and ritual use in the mountains of Oaxaca had been discovered by Gordon and Valentina Wasson in the summer of 1953. Of the many species which were in use in Oaxaca, subsequent laboratory tests revealed that only one species was easily grown and able to fruit under a variety of artificial conditions. That one species is* Strophuria cubensis—*the starborn magic mushroom. This book is a path to this mushroom; how to grow it and how to place it in your life like the shining light that it is.*[45]

The McKennas' book made the rounds in a growing community of trip-seekers. Almost twenty years after Wasson had brought word of "god's flesh" back from Mexico, closets throughout the country gradually filled with blue-bruising mushrooms.[46] Two years later, mycologist Steven Pollack released *Magic Mushroom Cultivation*, but his contribution to the community was not to go any further than that, as he was murdered in 1981 under mysterious circumstances.[47]

In 1983, Paul Stamets and mycologist Jeff Chilton were both inspired by the McKenna and Pollack books to pen their own guides.[48] One of them in particular, *The Mushroom Cultivator*, would go on to be the go-to *Psilocybe* cultivation manual, even though it wasn't explicitly about *Psilocybe* species. Mixed in with notes on the psychoactive mushrooms were instructions for growing culinary varieties; perhaps a nod to the growing range of species one could grow, perhaps a hedge against unwanted attention. When asked for advice on where to acquire *Psilocybe* mushrooms, Stamets has made a habit of answering, "Nature provides, I don't." Nevertheless, as he has also stated publicly, the book was designed to help those looking to grow *Psilocybe cubensis* as well as other mainstay "magic" species. It went on to become a foundational text for the home cultivation community, the handy "blue book" that served for many years as a point of reference for growers of all stripes.

Less than ten years later, another momentous text for *Psilocybe* mushrooms landed on doorsteps across the country, albeit far fewer than Wasson's 1957 article in *Life* magazine. It took the form of a classified ad in the September 1991 issue of *High Times*, posted by Robert McPherson, who went by the pseudonym of Psylocybe Fanaticus. Offered in the ad were some basic instructions and materials for "PF Tek": a nonmedical syringe filled with *P. cubensis* spores and a specially formulated nutritional substrate in which they could be grown easily inside glass mason jars. Though the mushrooms themselves were most certainly illegal, technically, selling the spores of the mushrooms was not.[49] Unsurprisingly, the FBI nevertheless took interest in McPherson, but he continued to run his one-sixth-page ads in the magazine for years until the DEA finally raided and shut down his business in 2003, landing Fanaticus in house arrest for six months.[50]

The kits were rendered largely unnecessary anyway, thanks to the next big step in the evolution of mushroom subculture and cultivation: the internet. As early as 1992, the details of PF Tek were widely available as a downloadable PDF. More and more people were growing the mushrooms, too, so spores weren't hard to come by. All that left was basic materials like glass jars, brown rice flour, vermiculite, and a closet or a similarly dark, discreet growing space. In theory, once a cultivator got started, they could carry on indefinitely; the mushrooms would simply produce more spores that would create mushrooms that then produced still more. This "teach a psychonaut to fish" approach might not sound like a sustainable business model, but despite this, and the involvement of many copycats, McPherson reportedly made $30,000 a month at the peak of his business.[51]

The worlds of psychedelic subculture, and especially those members of it that grew illicit mushrooms, stayed largely underground, gradually becoming more visible as the internet grew. *Psilocybe* mushrooms and the internet both carried socially transformative potential, but their associated cultures also shared real overlap. Former hippies turned professional and moved to Silicon Valley, kicking off the computer industry's "disruptive" ascent, to borrow from its own vernacular. Many were driven by a utopian vision inspired in no small way by the psychedelic counterculture of the previous decade; even Steve Jobs was known to experiment regularly with psychedelics.[52] In 2020, "microdosing" psilocybin is and has been de rigueur throughout the technology industry.[53] This connection makes sense. Among the counterculture, ravers, hackers, cybernetics geeks, virtual reality innovators and others found deep philosophical resonance in psychedelics, often drawing directly from the thinking of figures like Terence McKenna.[54] Humanity was seen as approaching some kind of zero point, *Homo sapiens* intellect having manifested the means of realizing at global scale the decentralized systems perceived in computer networks, psychedelic experiences, and nature itself. Digital technology was seen as "essentially a psychedelic substrate," as media theoretician Douglas Rushkoff puts it, the power of which was ultimately seized upon in service of capital, which could be said to have led us into a collective bad trip.[55]

In an echo of the open-source, system-hacking ethos that characterized many of the earliest communities that engaged with the internet and its precursors, the innovations of *Psilocybe* cultivators were often freely shared and improved upon online. A cornucopia of teks proliferated among the forums and message boards of sites like the Shroomery and Mycotopia. These sites are still active and remain host to vibrant communities. Many a mushroom cultivator has started their journey by trawling the seemingly endless forums, some comprising tens of thousands of threads filled with questions and answers, pictures from successful and failed experiments, along with a steady dose of philosophizing and drama. As new features and online platforms have emerged, the online mushroom community—which calls itself the OMC—has evolved along with it, developing its own cast of influencers, and an economy of trade and exchange, a distinct and open culture built around activities and experiences that were once discussed only in whispers.[56]

Mycotopia founder Hippie3 is one of the most influential, perhaps just as much as Fanaticus with his PF Tek. In 2005, the charismatic and mysterious cultivator known to most only by his online persona developed and popularized improvements upon the PF Tek recipe that went on to become the standard for many home growers. The implications of

Hippie3's next contribution to the OMC, the "airport," extended beyond psychedelic species of mushrooms, though. Many mark it as among the most important developments for DIY mushroom cultivation.[57]

Airport tek involves what is called liquid culture, a broth of nutrients that sustain countless suspended, living fragments of mycelium. The technique is useful for essentially any commonly cultivated species. No matter what kind of mushrooms you want to grow, contamination is one of the most vexing obstacles; a spore or two from a competing mushroom, or a few cells of bacteria, and suddenly there's competition for the hyper-rich nutritional environment of liquid culture, often abbreviated as LC. The solution that Hippie3 developed to address this problem turned out to be quite simple: drill two holes into the lid of a jar—one filled with breathable fabric, the other with a glob of heat-tolerant silicone—so that a syringe can be used both to inoculate and draw from the fungal broth with little to no risk of contamination.[58] Unlike working with spores or spawn, liquid culture allows for what amounts to instant inoculation, because it contains already-living mycelium eager to eat. This makes it possible to soak a substrate in active mycelium, its many eager hyphal tips reaching to meet one another and quickly forming a formidable mycelium.

Teks like the airport, along with other examples of low-tech or no-tech innovations in "guerrilla mycology" have made it possible for mass experimentation with fungi's many abilities in food, medicine, remediation, and beyond. Liquid culture, combined with something as simple as a plastic tub with two holes cut in it—called the still air box—make it possible for citizen scientists and cultivators alike to work with sensitive fungal strains in nonsterile environments.

Thanks to the spread of ever more accessible cultivation techniques, *Psilocybe cubensis* may be one of the most successful mushrooms the world has ever seen, in terms of pure profligacy. But the development of accessible cultivation techniques has also, appropriately enough, led to a proliferation of new ways of working with fungi, and many different varieties. What started as an effort to grow illicit mushrooms for personal use has evolved into the basis of new economic opportunities throughout North America, primarily based in growing mushrooms for food and medicine. Gradually, mushroom cultivation is engaging culture on a much broader scale.

CHAPTER SIX

Spawn Points

D own a short side street, a stone's throw from the Brooklyn Navy
Yard, Smallhold occupies a slender three-story building. Passing through its lofted offices and beyond the twentysomethings
busy at their computer stations, workbenches, and 3D printers, it called
to mind the offices of a youth media company, or perhaps a technology
start-up of the sort that had colonized much of postindustrial Brooklyn.
In a sense, both were true, but this start-up was unique in at least one
way: at the center of all the accoutrements signaling the "disruptive" and
cutting edge were the soft, curvaceous forms of mushrooms.

Smallhold's cofounder and CEO, Andrew Carter, led me into the
garage-turned-workshop where stacks of silver metal boxes lined with
long glass windows glowed violet. The wavelengths of the LEDs fluxed
throughout the day, to simulate diurnal cycles and stimulate the growth of
the pink, white, and yellow mushrooms. Cameras monitored their growth
in real time, while sensors and bespoke computer software regulated the
temperature and humidity of their environment. The boxes looked fit for a
long-term voyage in the hold of a spaceship, but in truth they were bound
for various restaurants and grocery stores around the city, where they would
be remotely monitored from the company's home base. As we started up
the stairs, Carter explained that it was all part of a distributed mushroom
farm, meant to enable its customers to grow and sell fresh mushrooms
without facing steep learning curves or unfamiliar distribution schemes.

At a handful of hip local restaurants and even a couple Whole Foods locations, one could hardly help but spot the company's eye-catching, cyberpunk "mini farms," the blocks of substrate inside bathed in deep-hued light and sprouting healthy clusters of mushrooms. "It looks all high tech, and there is a lot of technology and stuff, but we don't want to be unattainable," said Carter as we settled into seats on the rooftop patio across from the Manhattan skyline. "That's one of the things with mushrooms in general, they seem kind of unattainable because people don't know anything about them."

Indeed, it's difficult to guess your way into successfully cultivating and growing mushrooms. When it comes to growing vegetables or herbs, most people at least have some frame of reference for the basic elements and steps. It's possible to make mistakes along the way and still wind up with a delicious crop. That's less the case with mushrooms, where considerations such as sterilization and nutritional and environmental conditions are necessary to achieve reliable commercially viable results.

When I grew mushrooms for the first time, it required me to drive across Brooklyn to buy bales of straw from an equestrian center; find sacks and hydrated lime to sterilize the straw; order spawn to inoculate it as well as purpose-made bags to "run" the mycelium before the mushrooms start to grow—and I took one of the lowest-tech approaches. The least technical approach I ever took involved drilling holes into logs and stuffing them with shiitake spawn with a kind of specialized turkey baster, using melted candlewax to seal the gouges before stacking the logs outside and leaving them for the months-long incubation in the shade of the woods. Neither approach was very sophisticated, but like anything, they required time, attention, and trial and error.

Smallhold's idea is to take the preparation and guesswork out of the process for restaurants, grocery stores, and anyone else looking to add mushrooms to their offerings or diets. Each of the glowing boxes can produce between thirty and fifty pounds per week; the mushroom-bearing blocks within are replenished from HQ on a biweekly basis, enough for one "flush" of mushrooms. Many species will produce multiple flushes, but the weekly pace of replacement means there's not enough time for pests to show up. For example, *Trichoderma* is an especially common green, splotchy mold familiar to all mushroom growers. A little bit of *Trich* is pretty much harmless, but if you're hoping customers will ask "How much is that mushroom in the window?" aesthetics matter. "You can't have green mold, even though it's fine," said Carter. "A lot of big farms have mold everywhere, because nobody's going in there."

The company, which Carter cofounded with his business partner Adam DiMartino, emerged at an opportune time, rolling out its photogenic mushroom chambers and attention-grabbing business model right as mushrooms began attracting widespread public interest. Demand at restaurants and markets for specialty mushrooms had been consistently rising alongside interest in new models of local, sustainable agriculture, a bona fide consumer trend; as of 2019, according to the Institute of Food Technologists, "Vertical greenhouses, urban farms, and hydroponic gardens are poised to be the next wave in the local food movement."[1]

Alongside putting mushrooms in front of the growing numbers of consumers on the lookout for them, Smallhold's goal is to educate people on different fungal varieties, their culinary and nutritional potential, and basic ways of life. "When you talk to your average consumer, people don't know any of these things," said Carter. Prior to the pandemic, which remains in full swing at the time of this writing, the company regularly set up tables around various markets to pass out information and hosted tours of its facilities, all of which helped get the word out about the brand while also gradually demystifying its unconventional crops to potential customers.

Aesthetics also play a role in the company's success. Many mushrooms are naturally photogenic, but paired with the futuristic look of Smallhold's shiny flagship product, it's proved a winning match. A number of chic restaurants in New York City integrated the mini-farms straight into their interior design.[2] When fresh produce is at a premium, it's hard to do better than seeing it plucked from directly overhead as you sip your cocktail.

Despite its small size, the company quickly developed a prominent profile on social media, with more than eleven thousand followers on Instagram at the time of my visit. Smallhold easily caught the attention and interest of Brooklyn arts and media types, racking up numerous profiles in hip online magazines. "New York's Hottest Food Trend," its website read, atop a list of some of the most recognizable brands in digital media. Its mushrooms were the stars of a popular series of vibrantly colored photo books called *Mushrooms & Friends*, by artist Phyllis Ma.[3] "It's very trendy now, and I didn't see that one coming," said Carter, whose previous work saw him working with microgreen, hydroponic, and vertical farming outfits in and around New York City. "One of our first press things was in *Vogue* magazine.[4] I literally spent the last ten years in greenhouses. Like, it's not *Vogue* stuff."

Smallhold began its life in a converted shipping container on the Williamsburg waterfront, earning the first—and to date only—organic farm certification in New York City. That certification is complicated by the nature of the "farm"; the mushroom blocks are all prepared at Smallhold HQ, growing in identical climate-controlled boxes, operated all over the city and by a variety of customers. Arguably, the product is hyperlocal, harvested inside or right next door to the store that sells it, but that designation gets complicated, too, since the fungi and the substrate in which they grow are sourced from well outside the city. The vast majority of mushrooms sold throughout the United States are common *Agaricus* or "button" mushrooms, grown by the millions of pounds in Chester County, Pennsylvania. There's little room for anyone to break into that market, so Smallhold aims for the gaps created by a growing interest in specialty varieties, and those with the bandwidth and consumer demand to justify growing them.

"We saw an opportunity in mushrooms; anyone else can see that they have restaurants that want mushrooms but can't buy them," Carter explained. "We're inventing things because we have to, as far as technology, but the concept's not new. Like if you were to start a cherry orchard, you wouldn't go plant a bunch of cherry seeds. Nobody does that. They go and buy a bunch of grafted saplings from a nursery that just does that. A lot of tomato farms don't grow from seeds, they get grafted seedlings. I used to work on a farm that shipped them in from the Netherlands, I used to pick them up from JFK."

Smallhold's idea was never meant to begin and end with mushrooms. Originally, the vision was to distribute climate-controlled pods for growing microgreens, but it turned out that mushrooms made an ideal match for the model. As crops they're robust, highly nutritious, and a viable alternative protein source to meat; they're curious and delicious, with a growing niche market appeal that makes them a sensible side business or supplemental offering, hence the company's name. By creating a modular system that meets demand where it's at, and adapts to local environmental conditions, Smallhold also aims to minimize the waste of traditional agricultural distribution chains. Its latest line of store-shelf packaging are simple cardboard boxes with open windows through which customers can see and touch the mushrooms. In the future, they see potential for expanding the model nationwide, via a hub-and-spoke style network and centralized "macrofarms." One doesn't need to live in an area with a climate that's good for mushrooms to adopt their model, just space for the box and, of course, proximity to Smallhold HQ.

"What I think we can do is use certain aspects of food distribution in essentially shipping the living process," said Carter. "Then you grow right next to the customer, be that in a greenhouse, inside a store, next to a store. It can be wherever but as long as it's right next to them, then you get the best part of it, which is the freshness and the local aspect, but then you don't have to waste the food in the shipping."

One of the most commonly cited benefits of mushroom farming is food security. They grow on agricultural waste products such as straw, grains, and sawdust, and require little in the way of space to produce significant amounts of nutritional value. Just a few weeks after my visit, SARS-CoV-2 provided an unexpected test case for the company to prove this point, as restaurants shut down and grocery stores struggled to adapt to new pandemic restrictions. Facing questions about its role in a moment of pronounced food scarcity, and with investors concerned about what the new circumstances meant for the company's financial prospects, Smallhold moved quickly to adapt. Over the course of a whirlwind two-day planning session, it retooled operations to forgo the emphasis on mini-farms, instead focusing on sending individual mushroom ready-to-grow kits to customers' doorsteps.[5] Rather than monitoring and servicing a network of grow-pods, they would use their own growing capacity to sell mushrooms from a mobile shop, and deliver grow kits door-to-door. The latter took the form of bags filled with mushroom-growing medium, wearing cute HELLO, MY NAME IS stickers. Apartment-bound Brooklynites could bring the bags straight into their kitchens, looking forward to picking and eating the mushrooms in a matter of days.

It was a hit. "We have people quite literally living off our mushrooms," wrote Carter in an online article explaining the transition.[6] "This is a test to show how a technology enabled facility can be nimble and resilient in the face of a global pandemic, and can feed the world, one mushroom at a time." In the days when going to the supermarket was a tense and risky affair, and people were openly discussing a revival of the Depression-era Victory Gardens, the idea seemed right for the moment. A single Instagram post announcing the new initiative brought aboard five hundred new customers. The company ramped up production of mushrooms and grow bags, selling out of each on a daily basis.

The visual appeal of mushrooms paid off again as newly minted home-growers proudly shared the fruit of their labor (which really just amounted to occasionally misting the bags with water) on social media. Pictures of home-grown mushrooms flooded Instagram, festooned with digital googly-eyes and accompanied by excited testimonials; somehow the phrase *glub glub* became a meme in these posts, perhaps mimicking the sound one imagines the mushrooms make as they drink in air and burst forth from their bags. The context of the COVID-19 outbreak seemed to give the trending interest in mushrooms some new and unexpected direction, along with a bit of levity in an otherwise stressful situation.

The mushrooms themselves required little effort on the part of their new caretakers, their novelty and nutritional value welcome at a time when store shelves seemed spare and spirits in need of a lift. Those going stir-crazy in their apartments seemed to welcome a new hobby that also happened to feed them, while satisfying a similar caretaker urge as a houseplant might, or a pet, at least until it came time to eat them. The success of the approach suggests the role of mushroom cultivation for future food security and local agriculture, even—maybe especially—in dense urban settings. But mushroom cultivation has been undergoing a renaissance throughout the United States, in towns of all sorts and sizes, and often in far less flashy fashion.

The world of mushroom cultivation was first revealed to me at Smug-town Mushrooms, Olga Tzogas's scrappy company based in a Rochester warehouse. Its labyrinthine hallways, lined with living walls of flourishing mushrooms, were every bit as captivating and intriguing to me as the fungally abundant forest of our foray beside Lake Ontario.

Fans and plastic ducting regulated airflow to maximize cap size—mushrooms, like animals, absorb oxygen and dispense carbon dioxide. Too much of the latter, and the mushrooms will be "stemmy" and tough as they reach for oxygen-rich air, just as they do from the ground or the side of a tree. Ample oxygen leads to the wide, fanning caps that look and taste (and sell) the best. The ideal air mixes, moisture levels, and temperature ranges fit for most commonly cultivated mushrooms are well understood and easy to find, meaning that even beginners can produce hefty, healthy clusters with modest equipment and a little due diligence. Tzogas was no newbie, though, and under her guidance, the mushrooms flared from the sides of their plastic bags, marbled with white mycelia save for the coffee-colored flecks of undigested substrate. The bags for reishi and shiitake had their tops cut off, the burgundy and brown mushrooms stretching upward from their substrate in a valiant but vain effort to reproduce. The lifecycles of these mushrooms were confined and strictly regulated by human hands.

Dazzled though I was by Smugtown's output, at about two hundred pounds a week, the scale of Tzogas's operation would be considered modest by the standards of many specialty mushroom cultivators. A specialty mushroom is essentially anything but an *Agaricus bisporus*, the common button or portobello, as common to the produce aisle as broccoli or carrots. Any specialty mushroom grower is going to be downright minuscule compared to the massive, acres-spanning farms from which *A. bisporus* ship by the millions of pounds every year.[7] Sawtooth-topped *Agaricus* grow houses stipple the rolling landscape of Kennett Square, Pennsylvania, dubbed the Mushroom Capital of the World, though it's more like the mushroom capital of the country; China produces and consumes vastly more mushrooms than the United States, more than thirty-five million *tons* a year, upward of 75 percent of global production.[8]

Helpful to smaller operations is the fact that specialty mushrooms are not very shelf stable. Outside of a freezer, the life span of an oyster mushroom is just a few days, after which it starts wilting or falling prey to bacteria and insects, quickly taking on an unappetizing appearance. As with seafood, freshness is at a premium for specialty mushrooms so it makes economic and logistical sense for growers to operate locally. While shelf life puts a natural limit on the range of fresh mushroom sales, inoculated jars and blocks for home growers, tinctures and extracts and other value-added products are much less restricted. Many people find their way into mushroom cultivation because it is such an adaptable business; you could grow shiitake in some logs under the trees in your backyard; you could feed your family with a couple shelves of oyster mushrooms in a basement; you can grow medicinal reishi for personal use in your closet; you could convert a barn and grow hundreds of pounds a week of any of the above to sell to restaurants or markets. The limits are set by cost and level of commitment—as well as demand, of course—but as a rule, pretty much everybody starts small.

Locally sourced takes on particular meaning with mushrooms, too. The strains that Tzogas grows came from specimens picked straight from the woods; her prized reishi were derived from a mushroom picked at a city park in Rochester. Many mushroom growers do the same, preferring to work with strains and species that are native to their area. Properly managed, a sample from a single mushroom found in the woods can, and often does, become the basis for an entire product line. But many cultivators also attest that mushrooms simply do better in environments and conditions to which they're already accustomed. "Honestly I find our stuff, that's indigenous to our region, to be more prolific," she told me.

Growing native mushrooms satisfies some of the ethical and market demand for local produce and reduces the risk of introducing an outside strain to the environment; fungi are fecund, after all, and spores are promiscuous. And even strains plucked from the nearby woods manage to get around. Cultivators regularly swap petri dishes at festivals or through various exchanges, including online. If they make and ship spawn from these strains, they can go anywhere that packages are delivered. At the same time, a range of standardized strains are also distributed all over the country and beyond, to growers who don't want to go through the trouble of operating a lab and simply want a culture they know they can rely on for their business to produce consistent yields of predictable size, texture, and taste.

Smugtown, which carries the nickname ascribed to Rochester by journalist G. Curtis Gerling, sticks to the local strains as a matter of principle. Tzogas prefers to note that her company operates on traditional Haudenosaunee land, "in the city of so-called Rochester" and from its corner of postindustrial urbanity, Smugtown is unusual in that it's a business almost in spite of itself. "We are part of the many open-source 'mycommunities' that spread the knowledge of the wild, sustainability, and sovereignty," reads the company's biography on its website. "We aren't here to patent strains, or own anyone's culture. We want to encourage citizen science, seed saving, spore taking and species cataloging for the good of the planet not for profit."

Such an attitude would create a competitive disadvantage if competition were the goal, as of course it is throughout the mushroom industry as in any other. Among wild mushroom pickers, for instance, income is determined by what can be found growing in the woods; they rarely reveal and often jealously defend their most productive spots, meaning it can be a dangerous, even sometimes deadly line of work.[9]

When it comes to growing mushrooms, though, propriety is less about guarding turf than cultivation and growing methods, substrate formulas and strains. Even small operations retain their trade secrets, all of which causes Tzogas to roll her eyes. "Generally I think it's because the process is so simple that they're afraid if people find out how simple it is, they're going to lose a part of their market," she told me. "It's just kind of silly, because the market just keeps getting bigger, and there's enough production going on to meet the needs of the market. If you ask pretty much any mushroom farmer how much good mushroom product is going to waste, it's very little."

Shroom to Grow

About three hundred miles southeast of Rochester, nestled up against the eastern edge of the Berkshires in South Deerfield, Massachusetts, Mycoterra Farm operates out of a former horse stable and training arena.

According to the company's founder, Julia Coffey, Mycoterra produces about 2,500 pounds of mushrooms a week, making it the largest mushroom farm in the state. But that doesn't mean it's a dominant force.

"We're approaching half capacity," Coffey said after ushering me in from the snow. Outside the front door were raised garden beds made from discarded shiitake blocks. "At six thousand pounds a week, we'll be less than two percent of market share for mushrooms consumed in Massachusetts, so that's a testament to how much is coming from Pennsylvania and overseas."

Coffey began her business nearly ten years earlier, at age thirty, out of a closet-sized space in her basement. She had moved back to Western Massachusetts after studying agriculture at Evergreen State College and a stint working at Paul Stamets's Washington mushroom farm, bringing back with her a rich base of knowledge and experience in mushroom cultivation. Coffey launched her company with an initial investment of $5,000, and its growth has remained constant ever since; she decided to buy the house she was renting, cut a hole through the wall, and expand further into the basement. After six months, the basement was full, so she built a pair of forty-eight-foot-long greenhouses behind the house; those greenhouses have since been moved into the converted horse arena, where she employs fourteen full-time employees. The five-acre mushroom farm I saw was considered version 3.0.

Tall in stature and direct in her speaking, Coffey carries the confidence of someone in complete command of her work, certainly a factor in her ability to thrive in a competitive industry dominated by much bigger players. "Partly it was spite, stubbornness," she told me. "I'm not going to name names, but a couple guys in the industry told me I wasn't going to be able to make it, so, to prove it. And basically it was a challenge to see how many rules I could break and be successful. I worked in a multimillion-dollar mushroom farm, a state-of-the-art facility. Then I had a pressure cooker and flow hood, and built a plastic tent in my basement and closet. But, yeah, I made it work."

Near almost any city in North America, you're bound to find a specialty mushroom farm that's younger than ten years old. The largest of these farms are, naturally, the best known: Mycopolitan in Philadelphia,

Pennsylvania; Mile High Fungi near Denver, Colorado; Hi-Fi Myco in Austin, Texas; Earth Angel Mushrooms near Saint Louis, Missouri. In garages, basements, attics, bathrooms, warehouses, converted buses, trailer trucks, people grow mushrooms at various volumes for personal use, to sell to local restaurants, at farmers markets, and among their communities. Part of the growth and proliferation of small- and medium-sized mushroom growers is down to strong communities of practice that have formed around cultivation, especially online, making it easy to connect with the knowledge and peer groups that can take anyone as far as they're willing to go.[10] The materials involved are easy enough to find: wood beams, plastic sheeting, fans and ducting, perhaps some simple electronic sensors, along with the straw or grains to feed the fungus.

Besides proving she could do it, Coffey was motivated to launch her own mushroom farm as a way of demonstrating what fungi are capable of. The shiitake-block garden beds out front were one part of that; nearby, windrows of discarded mushroom substrate were the precursor to an organic garden, and the mushrooms themselves were grown on agricultural waste—straw, sawdust, seed hulls—from nearby farms. "The goal here is to grow as much as possible of a diversity of products on-site, to serve as a fungal demonstration farm and show the benefits of, basically, our by-product, and the soil-building capacities of the mushrooms," Coffey explained. "I have quite a bit of support in the neighborhood, but I want them to envy my yard."

The education piece was emphasized as a central goal of the company, as I've found it is for most people working with fungi. Mycoterra's mushrooms appear at local food co-ops, farmers markets, and in various products as value-added ingredients. The business has clearly benefited from growing consumer awareness about fungi, but Coffey suggested that there was still a long way to go in her mission of breaking mycophobia. "It's still active in our culture," said Coffey. "I went to my daughter's school and taught classes on mushrooms, but it's often not even in public school curriculums, often it's, 'Ew, mushrooms.'

"I'll be at a farmers market and somebody with a baguette and a bottle of wine and a wheel of cheese tells me that she's allergic to fungi. I'm like, you realize that's in all of the food you're carrying in your shopping bag right now? Yeast? Mold? People will come out of nowhere to tell you, 'I'm not going to buy anything, because I don't like mushrooms.' Like, do you go up to the guy at the farmers market who's selling eggplant and tell him, 'I don't like eggplant'?"

Past a long row of wood-beamed stables repurposed as storage bays, Coffey had me don a hairnet and booties before she led the way to a room where substrate was inoculated. Aisle upon aisle of storage racks held bags, filled with what looked like damp wheat bran as the mycelia began busily digesting, pockets of white fuzz scattered throughout.

Substrate is fungus food, usually some mixture of grains or sawdust, often mixed with nutrients that help certain species grow faster and more consistently. Sterilized or pasteurized straw is the standard for species such as oysters, favored for and by beginners because they readily eat such a wide variety of substances. Other species like reishi and shiitake enjoy hardwood, so sawdust is commonly used in those cases. Certain mushrooms like maitake are much more picky about their diets and schedules, making them significantly more difficult to cultivate and grow; some have yet to be reliably decoded, especially mycorrhizal species like morels and truffles.[11] Cultivators are constantly experimenting to determine what mushrooms will respond to, and new recipes are tried all the time.

Recall the basic life cycle of fungi: Most mushrooms fulfill their purpose by deploying scores of spores. Those that land in suitable environments form exploratory hyphae, which then grow through any food that's available, ultimately forming a mycelial mat that oozes protective and digestive enzymes as it grows. Once sustenance runs low or conditions inspire it—some mushrooms respond to temperature shifts, like shiitakes, which require being "shocked" in cold water before fruiting—the mycelium will shift energy toward producing mushrooms, which produce more spores, and the cycle repeats. In the hands of most cultivators, only part of this cycle plays out, and under tightly controlled conditions. Rather than sowing spores, cultivators typically inoculate substrate with spawn—living mycelia, usually mixed in with growth boosters like nutrified grains. On a bench in the next cavernous chamber, Coffey's employees were busily mixing spawn with substrate and pouring them into plastic bags slightly smaller than a pillowcase.

Roughly speaking, spawn is to mushrooms what starters are to plants. Spawn often takes the form of a bag of grains run through and held together by living mycelia, but it can also take the form of fungus-inoculated furniture dowels to be hammered into a log, or liquid culture for the fastest application.[12] When using grain spawn to inoculate substrate, growers will break up the block and the mycelium right along with it. The fungus, understandably, is averse to being torn into shreds, and goes into a temporary dormancy, but it is far from dead. Each segment carries the potential to carry on as an entirely separate organism. The spawn is mixed with the sterilized substrate, and the resulting mix is divided, stuffed into still more plastic bags, and left to incubate. At Mycoterra, this was performed in a designated area where traffic was strictly limited and surfaces sterilized regularly to minimize contamination. Cultivators compulsively spray their hands, equipment, tables, and everything else within reach using 70 percent isopropyl alcohol.[13]

A fungus often grows whether one wants it to or not. Cultivating is largely about clearing a path for them to thrive, in contrast to the great effort that is put into *preventing* them from thriving, albeit less often with concern for the spread of delicious mushrooms than plant pathogens or moldy food. Wooden posts and telephone poles are coated in antifungal chemicals like creosote. Fungicide is routinely sprayed over crops.[14] The rise of refrigeration represented a globe-spanning mobilization against fungi just as much as bacteria.[15]

In the wild, fungi are just one among countless other microbes competing for food. But for cultivators, it doesn't make much sense to have mycelia fighting for their lives while also trying to produce mushrooms. So they sterilize the substrate of competing microorganisms, whether to be used for spawn or to grow mushrooms for picking. There are many ways to do this, depending on the scale and type of mushroom food. Common methods include soaking in calcium carbonate–infused water, or pasteurizing overnight in a barrel, or using stovetop pressure cookers. Larger, better-funded operations will use autoclaves, which are expensive and incredibly difficult to move. But such large vessels are usually the only sensible solution for operating at commercial volumes. When scaling up, though, any solution that increases speed and efficiency can also make problems worse; a single error can propagate quickly and widely.[16]

"We do grow inside year-round, but you can screw up one thing and take down a whole operation," Coffey told me as she took me past the facility's big green autoclave. "I lost maybe a hundred thousand dollars last year due to contamination, simple human error, and equipment failures, and it can take months before you even realize you have a problem. By that point you've lost thousands of dollars of inventory."

Once substrate is inoculated with spawn, the blocks are moved to a fruiting room for a couple of weeks to incubate in the cool, moist environment they prefer. There, the mycelial threads run through and consume it before committing their newfound vigor toward producing mushrooms. Holes or slits are then cut into the side of the bag, and sensing the flow of air, the mycelium produces clustered "pins," also called primordia. These are baby mushrooms, ready to burst forth from the gaps over the course of a few days, gulping in oxygen as they expand and rain spores on the grow room floor. Until that point, the fungus has remained safely sealed off from the outside world, save for a cloth patch that allows filtered airflow that blocks competing microorganisms.

Some farms use giant polytube cylinders, like oversized sausages (or strawsages, as they're sometimes called) slung from racks or hooks for massive flushes to be picked straight from their sides. After the first flush, the bags will often produce a second, and even a third or more, generating weaker yields each time. One key measure of growers' success (and pride) is called biological efficiency (BE), which compares the weight of substrate to the weight of the mushrooms it produces. A five-pound block of straw that produces five pounds of mushrooms over the course of its life has performed with a BE of 100 percent. In most cases that's measured over two or three flushes.

But in each flush, the fungi's reproductive organs are thwarted from propagating, instead becoming food for the humans who have lured them out. In those terms, it all sounds somewhat exploitative, but the fungi enjoy a good, if short, life of uninhibited gluttony. As soon as they've reached maturity, but no later, they're twisted off and placed in a box for sale, ideally bought and cooked within the next few days. (The difference between a fresh mushroom and an old one is immediately apparent.) After two or three flushes, the blocks are usually considered spent. The plastic bags go in the trash,[17] and the spent substrate, along with the living fungus woven within, are often tossed onto a compost pile, as spent mushroom substrate makes for fantastic compost.

Spawn bags are made with special polypropylene that is rated to withstand the high temperatures of sterilization, but otherwise they're the same as the bags used to incubate and fruit mushrooms, complete with a little breathing patch. With some species, you can simply cut the spawn bag open just like a grow bag, releasing the sweet smell of living mycelium, the influx of oxygen signaling to it that the time has come to pin. But that would be a waste. With a typical ratio of 1 part spawn to 10 parts raw substrate, just one five-pound bag of spawn, properly distributed, could directly inoculate some fifty pounds of equally productive grow bags. This is actually why I took a stab at growing mushrooms the first time; frugal as I am, it bothered me to spend thirty dollars for a bag that produced about five pounds of mushrooms, so I used it to inoculate ten other bags in order to produce fifty pounds of mushrooms. (Well, it *could* have been fifty, if I had been more experienced.) It's this potential for scaling up that makes mushroom growing a viable, flexible business, at least on paper.

"It's really hard to grow mushrooms profitably," said Coffey, who half-jokingly lamented not going instead into microgreens, a direction she considered lower-risk than fungi. If your baby greens don't germinate, you can plant a few more row feet and till it in. "The numbers look good on mushroom farming so people think it's fun and profitable and, like, they can quit their programming job and get rich quick, but there's how things look on paper and then there's how things work in real life." At Myco-terra, for example, substrate demands for shiitake included producing nine hundred blocks of supplemented sawdust per week. At five pounds each, Coffee estimated they'd each be moved about ten times throughout their life cycle, making nine thousand reps for shiitake substrate alone. That doesn't include pushing and pulling them around on racks, moving other ingredients, or harvesting the mushrooms themselves.

In the former equestrian facility, there is plenty of room for the cultivation cycle, all pivoting around the sterilizing action of the massive green autoclave. Just beyond the "dirty" side of the tube, a loading dock with a big orange tractor is parked atop a pile of sawdust. That sawdust and other nutrients are mixed into a single batch, the moisture content set to about 60 percent, then deposited into the sterilizer bags, which are set onto carts and wheeled into the open mouth of the autoclave. Inside, steam builds up to a pressure of twenty pounds per square inch for four and a half hours.

Separated by a wall, the other side of the autoclave extends into a gleaming white staging room, where the sterilized bags come out the other side, to be unloaded and carted over to the inoculation table. From there, it's a short walk to the incubation racks, a long room with narrow walkways formed between the rows of sweating bags undergoing incubation. Taking quick stock of the aisles, Coffey estimated they held about a hundred thousand bags of different mushrooms at various stages of growth. After a couple weeks, they would move to the fruiting chambers, which include the two greenhouses from Coffey's backyard, rebuilt inside the equestrian center right next to the tractor and sawdust pile. Since the mycelium has already dominated the limited environment inside its bag, by the time the sides are sliced open, fruiting mushrooms don't invite the same degree of concern about contamination as inoculating or incubating.

Once fruited, the path to the consumer is generally short in terms of both distance and time. There is also a rich secondary market for mushrooms. In an on-site test kitchen, Coffey experiments to develop various value-added products with the mushrooms she grows, such as mixes of dried mushrooms and greens for soups. Unlike some growers, she doesn't have much interest in producing the medicinal tinctures or extracts that are quickly growing in popularity.

"The nutraceutical part of the industry is really cutthroat and kind of nasty, it's very competitive, and a lot of it is marketing BS," Coffey told me. "I believe wholly in the nutritional and health benefits of mushrooms, but I don't have any interest in getting mixed up in that mess. Like, I'm a mom, I run a business, I'm busy all the time. The products on the market that excite me most are the whole protein snacks, or the things that are easy and taste good and heal you at the same time. I'm resisting 'cyborg nation,' where we just pop nutritional tablets, and we're very embedded in the farmers market community, so I see opportunity for capturing a lot of waste streams and turning them into great products."

One of the most compelling virtues of growing mushrooms is that they can convert agricultural waste into nutrition and economic value. Spent grains, straw, seed hulls, and other by-products of agriculture can be turned into food for fungi, which in turn produces fungi for food and medicine. What's left behind are excellent makings for compost, which can grow still more food, enrich soil, or be put to use growing "secondary decomposer" mushrooms like *A. bisporus*, if a grower feels inclined to mix the mushroom compost with manure. Coffey wasn't interested in that, either. "I don't think they're really that good," she said of button mushrooms with a laugh. "They're overrepresented, and in order to produce them on an economic scale, you need probably square miles in buildings instead of square acres."

As we spoke, Mycoterra seemed to have reached the scale appropriate for its needs, with room to grow. But in reaching that point, Coffey suspected that she had not been working just in behalf of herself, her family, customers, and community. In parallel with growing her business, after all, she had been propagating fungi, the goal of the mushrooms themselves. "I think there is something to the sentient nature of fungi, they're driving us," Coffey told me at the end of our tour, invoking Michael Pollan's *The Botany of Desire*. "We think we're farming the corn, but really the corn is farming us; I don't run the farm, it runs me."

On the industrial outskirts of Denver, it was difficult at first to discern the facilities of Front Range Fungi. Surrounded by the sun-bleached jumbles of concrete works and the sinews of rail lines bundling up into the freight yards, it was hardly the landscape evoked by the company's name. The gourmet-mushroom growing facility had been forced to relocate several times since its founding in 2016, which for most businesses would mean tearing down, packing up, and starting over in a new building. For FRF, it meant hiring a driver, since the entire operation was contained inside a pair of semitruck trailers. Inconspicuously, almost invisibly parked in a dirt lot alongside a defunct factory, the only clue that I'd reached my destination was the heaping pile of hundreds of spent substrate blocks, dozens of mushrooms sprouting defiantly from them in the noonday sun.

The trailers stood with their back ends open with a raised platform connecting them, on which Trevor Garofano was cleaning out pressure cookers using a water hose. We had barely made our introductions before a pair of police officers walked in past the gate. They were polite, but got straight to the questions about just what *sort* of mushrooms this unconventional outfit dealt in. Garofano, an easygoing but savvy guy in his twenties, greeted the lead officer by name, and it was immediately clear that these visits were something of a routine. "So then the mushrooms, are they being sold to, like, restaurants?" the officer asked. "When we first got the complaint, it was that you guys were doing . . . *magic* mushrooms, since it got approved in Denver."[18] The young mushroom farmer was unfazed. "Oh, no. It's not really worth it, in terms of the law," he said.

After a few more questions about the company's plans for moving, the police left. Garofano, who had been with the movable farm for less than six months, stayed at ease and friendly throughout an encounter that might have left other people sweating. There was, after all, nothing to hide—inside the trailers grew various kinds of culinary mushrooms, the only known side effect of which was a full stomach. But the probing questions were secondary to the main reason for their visit anyway. Local building codes disagreed with FRF's presence in an industrial zone, and pressure was mounting for them to relocate. The company had only recently secured a new location about twenty minutes outside of town.

"Doing these smaller operations, you kind of run into that kind of stuff; you don't have the money, or you don't know every single little rule," said Garofano after the officers were long gone, satisfied that their message had been conveyed. "There's not a lot of rules for zoning, as she was saying, or regulations on mushroom growing, so that itself makes it fringe work. Like the same rules that would apply to growing lettuce or beans; they don't have that infrastructure yet for mushrooms. But yeah, at the end of the month, we're out of here. I know where we're moving, but I'm not telling them, because it's none of their business, really. Well, it *is* their business, but . . ."

Moving was a lot of hassle even for this relatively small operation. The pair of trailers represented the equivalent of what someone might set up in a small barn. Prior to the Denver location, FRF had been situated about an hour's drive to the south, in Castle Rock. With connections to the restaurants and food activism communities in and around Denver, they were quickly back up and running in their new spot, but it had meant a suspension of production, compounding the financial complications. "We were supposed to be moving literally last week, so we stopped shiitake production for a little bit, and then the move got pushed back, so now we're really low," said Garofano as we sat down in a small office space in one trailer. "As a business you want to be able to continuously have stuff. The product can't just be made and then stored; it only has like four days once I pick it before I can sell it, and anything after that is just waste."

Luckily, when it comes to mushrooms, even the waste isn't wasted. Outside, the pile of spent substrate blocks were still producing well past the typical one or two prime flushes that a farm will pick and sell. Given time, and removed from its plastic, the blocks will transition into soil, even producing mushrooms along the way that can be picked and eaten or sold. "A lot of the stuff we do is pretty self-contained," said Garofano. "We don't waste a ton, in theory. A lot of those blocks get put into someone's yard, and they fruit."

Just a few days prior, our mutual friend Mercedes Perez Whitman had swung by to swoop up some of FRF's spent blocks. She did this often; the nearly endless stream of commercially nonviable blocks were still perfectly good for growing food, so she buried them in the garden around the quarter-acre housing collective where she lived. Her landscaping business, MycoSprings, did the same for clients in and around Colorado Springs, offering a source of food, potential income, and fortifying soil in the process. After my visit with FRF, I drove down to see it for myself. Around the edges of the house, firm, dark gray little tufts of oyster mushrooms sprouted from the floor of the garden beds, which were picked, cleaned, weighed, and set neatly in little boxes at a farmers market across from city hall. At five dollars a box, she sold every one.

For FRF, the unusual arrangement offers flexibility, which comes with its share of frustrations and trade-offs. Leveraging a scrappy mentality and a clever system originally designed by systems engineer and cannabis entrepreneur Michael Ring, they've managed to make efficient use of limited space.[19] FRF uses an app called Almond to control the temperature and humidity of the farm's contained environment, and a network of interlinked wireless routers to monitor and control all manner of growing factors. Garofano pulled his phone out to show me. "Essentially, as long as this thing's got Wi-Fi, I can be at my house and control it," he explained. "I can control the lights, I can shut them on and off, I can control the humidity. There's heat, there's fresh air, there's a swamp pump that basically works with the swamp cooler to help us fluctuate the humidity."

One trailer was dedicated almost entirely to fruiting; in the back end sealed by sliding glass doors, cramped rows of shelving stretching down its length, multispectrum lights stimulated growth while ventilators kicked in periodically to fill the room with cool mist, all monitored and adjusted through the bespoke control system. Square holes had been cut in the adjacent sides of the trailers, and connected with a short chute so that fully incubated blocks could slide straight from the inoculation side to the fruiting side, one among many eked-out efficiencies. "We literally just cut sides out of the trailer and then just pass them through," chuckled Garofano. "It's just a little faster than bringing them all the way out here." The whole thing was "very bootstrap," as the team liked to say.

Not everything took place inside the trailers. Deployed on the ugly linoleum of the nearby building were two tall plastic hoppers used to mix substrates, often wood fibers and soybean hulls. A pair of grocery store coolers had been repurposed as steam sterilizers, used in tandem with the pressure cookers being cleaned outside. Neither could match the pressure or volume of an autoclave, which meant accepting a certain number of contaminated bags; so far, the margins justified saving on the expensive, but much more capacious and consistent pressure vessel. Next door was a room stacked with fiber-reinforced plastic sheets, the site of an aborted attempt to build out the lab externally. At least until they moved, it would have to stay in the trailer where all the substrate was inoculated, and where we sat talking.

The lab had a small refrigerator for maintaining cultures, next to a workbench with a HEPA filter for keeping contaminants out of petri dishes. Cultivating fungi requires particular skills and equipment that are distinct from producing spawn or growing mushrooms. Some companies undertake all three, but most opt to focus on one or two, the smallest often starting their production with spawn bought from elsewhere, or grow straight from the bags to simply pick and sell the mushrooms at a premium. But going through the trouble of maintaining a lab and culture library affords opportunities to select for and even breed unique strains that exhibit favorable properties: yields, flavor, color. That means a lot more control over the end product, and the possibility of making and selling spawn for others to grow the same mushrooms.[20]

The basic purpose of a cultivation lab is the isolation, maintenance, and propagation of fungal strains. That can take several forms. One of the most basic methods is cloning; that is, taking a small piece of the mushroom you want to reproduce, and using that to grow mycelium that then goes on to inoculate substrate and produce more of that mushroom. Every part of a fungus, mushroom and mycelium alike, is made of hyphae, which are "totipotent." That means every single cell can start a whole new organism, multiplying into mycelia that ultimately produce mushrooms. When a cultivator finds a mushroom to clone, they'll usually take a small core sample from deep inside, the same way you would if you wanted to sequence its DNA, and for the same reason: fewer competing microorganisms to complicate matters. When a tissue culture is placed on nutrient agar in a petri dish, the cultivator will have to be sure the organism that grows there is indeed the fungus they seek. Another organism could easily get onto the plate and confuse things, or create contamination that can be carried through to the spawn, hence the HEPA filter and the usual routine of spraying everything down in alcohol. Another technique is to collect the spores of a mushroom, and germinate them under controlled conditions so that they produce a fresh, genetically complete strain capable of producing mushrooms when mixed with the right substrate.

Once healthy mycelium has formed on the agar plate, a chunk can be removed to propagate. Often, a cultivator will remove their chunk from the edge of the mycelium, where the organism is reaching out and most vigorous. If done correctly, one little splotch is enough to launch an entire mushroom-growing operation; remove a section, drop it into grain or another nutrient-rich substrate, assuming proper sterile technique, and in a matter of a few weeks, it can be built up to production levels. Sometimes, though, a culture simply arrives already thriving in the petri dish, by way of purchase, trade, or donation from another cultivator. Growers will keep their own culture library—usually as stacks of marker-scrawled petri dishes in a cooler or small refrigerator—but for spawn makers, cultures are the foundation of what they do.

Garofano removed a petri dish with a little dream catcher of mycelium in agar, and explained the multistep process. A small amount of mycelium taken from the agar plate inoculates a small glass jar of substrate; once fully colonized, that small jar of spawn goes on to inoculate a larger spawn bag, which then inoculates ten or so fruiting bags.

In the meantime, the source of the living cultures has to be kept in a state of stasis, so that the fungus doesn't overexert itself or become less efficient over time, reliably reproducing whenever a small chunk is removed and used to start a new batch. If a strain becomes less vigorous over time—a natural process called senescence—it's possible to take genetically compatible spores or mycelia and breed them to produce a new strain with restored vigor.

Breeding a successful strain is a matter of selecting for desirable traits and exercising caution. For instance, two individual fungi of the same species could get into a culture and, if they're not genetically compatible for mating, simply stake out their share of available nutrients and basically cut in half the potential of a block or agar plate or, if unnoticed, a whole bag of substrate.

The basic tools of the trade are agar medium and petri dishes, a scalpel, flame source, and ideally a laminar flow hood and HEPA filter to move air while screening it of microbes and other contaminants so that they stay out of the culture. Combined with mastery of a few basic techniques, it's possible to grow mushrooms in many environments and at a variety of scales. At smaller scales, the risks and consequences of contamination are negligible compared with producing thousands of pounds a week.

For FRF, which like Smugtown was producing about five hundred pounds a week, restaurants were the main customers, making the delays created by relocation a cause of real stress. Restaurants aren't generally concerned about their suppliers' logistical woes; they simply expect their product. For Garofano, it meant working on the other side of his previous career. Having moved to Colorado from Massachusetts, he took the job at FRF with plenty of experience as a line cook but little in the way of mushroom-growing chops. Through an internship at The GrowHaus, a nearby community food nonprofit with its own small on-site mushroom farm, he learned the ropes and found himself becoming fond of working with mushrooms.

"I do have some sort of pride in it, I do like growing mushrooms; I like cooking and eating them, so it's nice to be able to pick them and bring them home," he told me. "It's a good way to work and make a living, but to me what's really fascinating about the fungi in general is, it's everywhere. It's everywhere under this ground. Everywhere you walk on the earth, it's there."

Sitting cross-legged on the floor, Willoughby Arevalo quizzed his three-year-old daughter, Uma, with a deck of mushroom cards. "Where is . . . the *blewit?*" he asked in a gentle voice, and after a moment Uma pointed straight to the correct card. "Where is the slippery jack? *Suillus?* Where is . . . the *destroying angel?*" Nearly each time, the child chose correctly. "She's steeped in it, what do you expect?" Arevalo said. It was to be expected, and partly, also a matter of safety. "We have death caps growing wild in the front yard."

The cozy first-floor apartment where Arevalo and his family lived in Vancouver, British Columbia, was partly underground, its interior surfaces covered with artwork and books, mostly related to mushrooms, ecology, and land stewardship. He and his partner, Isabelle Kirouac, were also collaborators as artists and educators, emphasizing productive and reverent engagement with nature. Various musical instruments and equipment were stacked in the corner of the living room, including a Roland brand amplifier, its face modified so that it read simply, LAND. A laptop on the counter featured a sticker with the word MYCELIUM rendered in the tangled, tortured font of a death metal band's logo.

After the mushroom card game, Arevalo and I sat at their dining table, where I learned about his history with mushrooms as he taught me how to forge them out of felt. From a picture book, I chose a *Gomphus*, somewhat reminiscent of a chanterelle; Arevalo chose an earth star, which looks just the way its name suggests. With barbed pins in hand, we spoke as the rough outlines of our chosen subjects took shape. (At the Vancouver Mycological Society meeting we attended, Arevalo presented a strikingly realistic felted death cap to its former president, Paul Kroeger, who laughed and gave it a playful kiss.)

Arevalo, his expressive face framed by a wild mop of curly hair and an unruly beard, had dedicated most of his thirty-six years to mycology, but not with a career in mind. "I studied art in university, and I took all the mycology classes," he told me, peering intently at the mushroom slowly forming in his hands. The lack of a degree in fungal studies, though, was not to be taken as a sign of his lack of expertise. Arevalo began studying mushrooms at age four, growing up near a very "mushroomy zone" of redwood forests in Arcata, California, which he duly identified as traditional Wiyot territory, adjacent to Yurok territory. By age nine he was regularly attending meetings of his local mushroom club, although he wasn't the most active member. "It was hard for me to drop into it and really follow it as a person of that age," he told me. "While the older generation that was prevalent in the mycological society was supportive, and welcoming, and warm, and excited to have an excited, interested kid who was starting to build a little knowledge base, they didn't really manage to keep me engaged."

Instead, he pursued an independent education, quickly gaining ahead-of-his-age experience in the identification and edibility of wild mushrooms. Born into a culture that encourages caution around mushrooms first and foremost, it would be years before he could put his skills to the test of actually eating what he found. On a trip to the beach at age ten, Arevalo spotted a downed alder with a cluster of what he thought to be oyster mushrooms growing from it. His father was actually quite fond of fungi and encouraged his young son's interest, but parental concern prevailed and the mushrooms went uneaten. "Looking back I know I was right, they were definitely oyster mushrooms, and I know my dad wanted to let me. *He* thought I was right, but he couldn't let me eat them." Gradually, Arevalo became interested in fungal ecology, and ecology in general. Eventually, his interest in fungi began to extend beyond the search for edible varieties. Reading the book *Mycelium Running* by Paul Stamets got him excited about their critical roles in ecosystems, the many applications and partnerships that were possible with fungi. Arevalo's path echoes that of many who compose the emerging fringe of mycological culture.

"I realized that my whole relationship up until that time with mushrooms was not a reciprocal one," he said. "I had gratitude, I did have that, and I would often talk to the mushrooms and thank them and things like that, but you know, I was *hunting* them, I was naming them and identifying them, and so I realized I was not honoring their agency, and I was objectifying them. That didn't sit well, and it was kind of a gradual realization, it took me some time to process that, and also to figure out what to do with that. And what I figured out was that I could grow mushrooms, and engage in regenerative cultivation practices to try to support the fungi that I loved through cultivation, and share what I knew."

At the time of my visit, Arevalo had recently released his first book, *DIY Mushroom Cultivation*, which outlined the means and methods by which almost anybody could grow mushrooms at home, with minimal resources. His own apartment was testament to that reality. In the water heater and furnace closet just off the kitchen, spawn was incubating in the darkness. Pulling a cardboard box from the top of the fridge, he selected mason jars one by one and held them up to the light to show me their contents. They contained liquid cultures, nutritive broth cloudy with pure mycelia of various species. In one, the filaments of lion's mane fungus had conspired to make a break for it, coagulating into a column of mycelium that reached up to the lid of the jar. On another, the words HAS BACTERIA were scrawled in big block letters. "I want to see what it'll do," Arevalo said with a shrug as he placed it back into the box. Each jar contained a marble, so that the cultures could be stirred regularly to break up the living mycelia inside. The mycelium could in this way be kept in a state of potential, without the need to ever open the jar; as doing so would inevitably contaminate the contents. These cultures were produced not in a sterile lab but in the family's kitchen. Arevalo is a modest master of cultivating mushrooms with limited materials and in unconventional environments. Grinning, he showed me a picture of a block of pink oyster mushrooms he fruited from the hanging soap rack in their shower.[21]

Arevalo and Kirouac relocated to Vancouver in 2013, moving into a one-hundred-square-foot cabin, next to what was described to me as a "chaotic clown house with thirteen people living in it." It was in this environment that Arevalo sharpened his methods of DIY mushroom cultivation, having previously spent years working at a large mushroom farm in California, where he honed his skills. The practice and sharing of these skills quickly became a driving passion. Such skills have also become central to an emerging generation of amateur mycologists, skills that are not often represented at the traditional mycological society meetings. At the typical mycological society meetings, cultivation is rarely a subject of conversation compared to the identification and consumption of wild mushrooms.

After becoming interested in the deeper possibilities posed by fungi, Arevalo pursued what felt for years like a solitary path, not thinking he'd ever find community around his desire to grow mushrooms in a thrifty and reciprocal spirit, or with an eye to their deeper meaning and potential. That is, until he saw a flyer for something called the Radical Mycology Convergence at the local food co-op. "My jaw dropped and I was like, what the fuck is this? Why don't I know about this? Like . . . *what*? Who's doing this?" It was 2012, he and Kirouac had just fallen in love, and suddenly he'd been made aware of a community of like-minded mycophiles that he didn't even know existed prior to encountering the flyer. "I was like, holy shit, there are other people like me. I'm not alone in this, and there's other people who know more than I do!"

Many people grow mushrooms because it's a potentially profitable enterprise; some enjoy the challenge of it, and of course some simply want to grow their own food and medicine. But whether for food, as medicine, or to initiate a consciousness-expanding experience, growing mushrooms is also a means of exercising agency and intention, for ourselves and our communities as well as in relation to another life-form. That is part of why cultivation practices sit at the center of mycological communities emerging throughout North America. In kitchen closets, basements, garages, barns, warehouses, or acres-spanning farms, growing mushrooms is tantamount to communion with them. It's a chance to exercise the reciprocity of which Arevalo spoke. The profile of fungi in our culture is gradually shifting from one of disgust or deviance to that of healing, partnership, and opportunity. That starts by learning how they work, and how to work with them.

CHAPTER SEVEN

Building a Myco Scene

Taking my exit for the Cascade Highway that crosses the rural hamlet of Mulino, Oregon, I still wasn't sure exactly where to find the Radical Mycology Convergence. Having missed the instructional email that went out to ticket-holders, I was relieved to receive a text message from the organizers, and just in time. Taking the next turn, I drove up and onto a gently sloping plateau topped by quilt squares of ranchland and neatly hemmed stands of pine. Shifting in soothing parallax beyond the trees, pleated foothills gathered toward the looming vertex of Mount Hood.

At a bend in the road, I spotted the entrance to Brown Bottle Farm, where a cheery parking attendant rose from her folding chair to guide me into a dirt lot, and parked me among dozens of vehicles hailing from various states. It was midmorning and attendees were just starting to trickle in, but clusters of tent domes had already sprouted up around the forty-acre farm. A bottleneck of visitors was forming to check in at the information desk, identified by a hand-painted sign as THE HYPHAL TIP. Every gathering spot, I quickly realized, was designated with a fungal sobriquet: THE SCARLET CUP, THE PAX, CAMPASPORA. The queue was a wood-chip-carpet runway show of wide-brimmed hats, hardy hiking shoes, and bulging backpacks festooned with canteens, carabiners, and reusable bowls. One sturdy, elderly man with a substantial beard and red work suspenders looked for all the world like an off-duty Santa Claus.

From somewhere downrange, the strains of Tuvan throat singing wafted along the breeze.

Just a few paces from the Tip, a volunteer-run kitchen was ramping up the daily lunch line; food was paid for with the ticket, but everyone was expected to bring their own plates and silverware, another memo I missed. Nearby was a small village of shade structures, where various mushroom-related products and paraphernalia were available to try or buy. I stepped under a canopy where books, stickers, and spawn bags competed for table space next to a list of other offerings: T-shirts, patches, dowels, kits, syringes, petri dishes, *Cordyceps* popsicles. In need of a pick-me-up, I bought a cup of *Cordyceps*-infused coffee, declining to have a pair of the stringy orange fruiting bodies tossed in ("without the worm," as they put it). Taking my first furtive sip—which, despite my slight trepidation, tasted like coffee—I noticed another vendor holding a dropper of tincture over the outstretched tongue of a man in a cowboy hat. In the shadow of a big red barn just downslope, a large and appreciative crowd had gathered to listen as a trio of performers took turns dramatically reciting a poem, told from the perspective of a fungus after the apocalypse. As I listened, a woman walked by with her children lagging behind. "Come along, little spores!" she chided. Suddenly feeling the need to get my bearings, I sought out the event's de facto center of gravity.

Peter McCoy was easy to spot. Tall, soft-spoken, and sporting a distinct ensemble of baseball cap, thick-rimmed spectacles, and tightly braided ponytails, he was made even more conspicuous by the car-tire-sized piñata he held, shaped like a split-gill mushroom. If radical mycology can be considered a brand, McCoy is its best-known ambassador, though he seems to demur at such notions. The activity swirling around us wasn't meant to be any one person, after all. Rather, the RMC was the momentary nexus of a distinct but loosely defined mycological community, an emergent "mycoculture" that had developed over the previous decade. Its animating ideas have circulated in zines, YouTube videos, books, community workshops, touring symposia, mix-tapes, an online fungi film festival, and convergences like this one, through which McCoy and his collaborators hoped to encourage a kind of fungally informed socio- and eco-consciousness.

"We don't do a lot of mushroom cooking classes, we also don't do a lot of mushroom foraging," McCoy said as he set down the piñata. "This event, this forum, attracts a narrower spectrum, I guess, of mycology people. But generally, when I teach, what stands out is the sociology, how it just draws in so many people of so many demographics, cultural backgrounds, ages, and reasons for caring about this."

The goal was to foster a respectful working relationship to fungi, founded on an appreciation of their importance as food, medicine, ecological keystones, potent allies, and resonant symbols. A special emphasis was placed on low-cost, accessible cultivation techniques. The infusion of gnostic mushroom history, permaculture sensibilities, citizen science, and the promise of seemingly alchemical feats—such as transfiguration of liquid culture into mushrooms, or training mycelia to eat used cigarette butts—made for an alluring, heady mixture.[1]

McCoy and I were speaking at the fifth RMC, a gathering that takes place every two years. With some five hundred people in attendance, it was the biggest convergence yet, signaling the broadening appeal of a socially engaged mycology. By coincidence, the North American Mycological Association's annual foray was also taking place about a half hour away in Salem; NAMA is the largest organization of amateur mycologists in the country, and some of the largest mushroom clubs in the country are NAMA affiliates, including one or two in the Pacific Northwest.[2]

"It's just in the air here, literally and figuratively," McCoy said. "And it's increasingly popular generally, culturally. One of the things we really try to do is to make it cool, that's very intentional. How can we make this not for middle-aged predominantly, and gray haired, really, in the mushroom clubs? Not that it's not for them, and not that I don't like them, or don't get along with older folks at all, but that's actually a noted problem. People kind of 'higher up,' if you will, in the organization of the mushroom clubs of North America, have asked how Radical Mycology can support them, because they need younger people, and I'm all for it."

A quick glance around was enough to see that the generational makeup of the group skewed much younger than the forays and conferences of the mycological societies. Attendees, speakers, and facilitators alike hailed from a variety of backgrounds. There were permaculture designers, homesteaders, students, educators, cannabis farmers, artists, and citizen scientists, representing a diversity of relationships to fungi and what they represent.

"This feels really different from every other mushroom event that you'd go to on the West Coast," said Christian Schwarz, whom I found sitting in the grass as young naturalists, loupes in hand, gathered around to get his help in ID'ing various strange things they'd uncovered in the nearby forest. The event, he said, catered to "younger people, more sort of punk, radical, focused on art, focused on community. I think people really like that, and it was not addressed for a long time."

The cultural emphasis was clearly visible all around us. Brown Bottle's big red barn had been converted into a fungal art exhibition; a stage was built in the forest-clearing-turned-auditorium, where hundreds gathered for a nighttime talent show; the final day featured the "first-ever myco-Olympics," in which teams with names like Bleeding Mycenas, Psychedelic Reindeer Piss, and The Fantastic Spore competed in gamified interpretations of mycological activities. Participants lined up at a table to race through a mock inoculation, or mimicked mycelial behavior by sprinting with straw-stuffed burlap sacks as McCoy, hoisting a microphone and a hefty speaker, took on the role of sports announcer. Kneeling beside a group of contestants furiously digging into buckets of red Jell-O, it occurred to me that I still wasn't exactly clear on just what radical mycology *is*.

McCoy outlines a definition at the outset of his massive tome of the same name:

> 1) *A social philosophy that describes cultural phenomena through a framework inspired by the unique qualities of fungal biology and ecology. 2) A mycocentric analysis of ecological relationships. 3) A grassroots movement that produces and distributes accessible mycological and fungal cultivation information to enhance the resilience of humans, their societies, and the environments they touch.*[3]

"I try to say that it's more of a philosophical approach," he said when I posed the question directly. "*Philosophy* is a heavy-handed word, but it's a way of thinking differently about this topic that has had a very limited presentation. What does it mean to just totally open it up and put it into modern contexts, whether those are social, political, spiritual, whatever else? Let's put fungi into these conversations we're having about all these other topics, because they're often not there, and what does that look like?"

The project of the RMC seemed to be no less than fostering a mycologically enlightened society. By spreading knowledge about what fungi are and how they live, by growing them for food and medicine, and benefiting from their ability to restore landscapes, dissolve waste, generate materials, among many other "applications," the hope is that we might come to shape a culture and way of being that reflects the reciprocal, regenerative ways of living so often exemplified by the fungi themselves. The mycelium is the message, as McCoy likes to say.[4]

At the center of this nascent mycoculture are the techniques and practices of DIY mushroom cultivation. Its roots—or hyphae—are in the underculture of *Psilocybe* cultivation teks, informed by the attitudes of the hacking and open-source technology communities. At the RMC, these influences found mycological expression in a Culture Barter Faire; a MycoHack inspired by FarmHack, an open-source community focused on agricultural technology; an event listed as Spores and Strains described itself as a "non-facilitated, non-commodified free-for-all where participants swap cultures of their favorite (legal) species."

The only course that McCoy himself taught was about DIY—or as the class title put it, "DIT" (for Do-It-Together)—mushroom cultivation. In a crowded, dimly lit basement room, he passed around spawn bags while flipping through projected slides demonstrating the basic vocabulary of low-tech mushroom cultivation. The Hippie3 airport lid for liquid cultures; "glove boxes" for handling cultures, made with plastic tubs and rubber dishwashing gloves; soaked cardboard (also known as "poor man's agar") as in-a-pinch substrate. These methods of growing mushrooms required only glass jars, coffee grounds, and other low-cost, easy-to-acquire materials. Beyond the opportunity these methods offer for almost anyone to grow mushrooms, the more people practicing them, the greater the chance for unexpected insights or advancements that might inspire or enable still more folks to engage with fungi in novel ways. A show of hands revealed that most people in the workshop were brand new to mushroom cultivation.

There was immense expertise gathered on Brown Bottle's grounds, but many, including those presenting talks, could be fairly described as amateur. One speaker pointed out that the root word of *amateur* is *amator*, which translates roughly to "doing something out of love." Whatever form the work took and at whatever level one felt moved or able to participate, there was a standing invitation to do so. "Meet people where they are," was a phrase that floated around like a kind of unofficial credo.

Radical Mycology could be understood as one facet of a broader movement, which ethnographer Dr. Joanna Steinhardt identifies as "DIY mycology," in which she discerned an "interspecies (or cross-kingdom) engagement that is part of an emergent ecological ethics and deep ecology worldview."[5] This outlook offers some amount of contrast with the more traditional forms of popular mycology, focused as they are on foraging and identification, as well as the culinary and, to a lesser extent, psychedelic dimensions of mushrooms. Perhaps epitomized by the North American mushroom societies that formed in the '60s and '70s, these communities emerged from the milieu of the hippies and back-to-the-land movements that largely fizzled off in the communes, or found new purpose and perspective in Silicon Valley.[6] In DIY mycology, there is the sense of a scrappy young subculture motivated by a righteous mandate to work with mushrooms to "save the world," only for real this time.

"When you look at Peter and the other Radical Mycology people, they're influenced by punk, they're influenced by anarchism, and there are these other threads that come out of that," said Steinhardt, whose research described a mycological movement that echoed ecological, political, and countercultural movements of the past decades, which rose in response to looming ecological disaster and unsustainable social disparity. "I think in the seventies there was this sense of a kind of urgency of how they were going to change the world, but I think that now the urgency is different. It's not like, 'We need to change the world because patriarchy is bad,' it's, 'We need to change the world because we might be extinct in one hundred years.' It's much more urgent."

In the worldview of DIY mycology, fungi are seen as powerful potential allies in addressing food shortages, erosion, soil and water contamination, providing medicine. But also wrapped up in working with fungi is the need to adjust our relationship to them and, by extension, the natural world more generally. Thus, we don't "use" fungi, but rather "partner with" them; we *meet* mushrooms in the woods; people had come to the convergence to "expand the mycelial network," and to "spread spores." Underlying all this, principles of emergence and systems thinking emphasized decentralized, nonhierarchical communities, kind of like mycelium itself, a comparison that didn't go unnoticed.[7] "That's always been the approach," said McCoy. "It's sort of tongue in cheek, but also sort of serious, the metaphor. It's a very good metaphor, when you think about it."

Seen through that lens, the convergence was an effort at inoculation analogous to what cultivators do in their labs. The spread of a fungus, after all, expresses distinct patterns, and perhaps the same can be said for human culture and society. Anyone with experience in fringy communities or subcultures—as part of a punk music scene, say, or in arts and activism spaces—would quickly recognize the organizing principles of DIY mycology and the Radical Mycology Convergence. Gathering at unlikely (often unsanctioned) venues, people commiserate and share knowledge, skills, food and drink; music and art and poetry and other forms of value- or meaning-making that serve to reify the sense of collective sentiment and purpose. At the RMC, that purpose was partly about seizing upon a seemingly unique moment in history.

"We are in such an unprecedented era of human-fungal relationships," McCoy told me, describing what he sees as a kind of "fourth-wave ethnomycology." "There was this vast span of history, of humans working with fungi all around the world that we know almost nothing about. Then there were the foundations of traditional mycology, roughly two hundred fifty years ago, a handful of people just studying, naming, and classifying fungi for a couple hundred years. Then there was the psychedelic counterculture in the middle of this last century, and that paved the way for new insights into fungi. And now we're in this whole new era, where we can talk about them not only ecologically, really insightfully, at greater depth, and with a more holistic understanding than we ever thought possible, but we can also cultivate them with ease, in a way that's never been possible in human history. And we can apply those cultivation skills in all these fundamental, profound ways, all these technologies and applications that are unprecedented. I think this is a major turning point. We're going to look back in three hundred years and see this as the start of something that, I hope, transforms into the mycoculture of the future."

———

Radical Mycology emerged in Olympia, Washington, out of conversations, independent publications, off-grid gatherings, and community projects that first took shape under the banner of the Olympia Mycelial Network.[8] The region offered an ideal substrate, as it were, for the germination and propagation of its mycosocial ideas.

The lush, misty state of Washington is the only one completely encompassed by the Cascadia bioregion, along with parts of Oregon, California, Idaho, Nevada, Wyoming, Montana, Alaska, Yukon, and British Columbia. "The delineation of a bioregion is defined through watersheds and ecoregions, with the belief that political boundaries should match ecological and cultural boundaries, and that culture stems from place," according to the Cascadia Illahee Department of Bioregional Affairs, an advocacy organization that plays the tongue-in-cheek role of an official agency.[9] If there is an ideal test case for the bioregionalism concept it is Cascadia, which, so defined, includes the entire watershed of the Columbia River, bound inland by the Cascade Range from Northern California up into Canada. More often the place is referred to as the Pacific Northwest. The Cascadian landscape and watersheds are profoundly rich in biodiversity, host to precious old growth forests, vital fishing, agriculture, and aquaculture that have supported the region's rich Indigenous cultures for centuries. It is estimated that First Peoples numbered over half a million in the region as of 1750, prior to European colonization.[10] Tribes such as the Nisqually, Duwamish, Squaxin and many others retain a significant cultural and economic presence in the region.

In Olympia, a hearty culture of social and food justice activism was bolstered by agricultural communities, independent publishing and zine scenes, and enhanced by the weirding influence of nearby

Evergreen State College. The anti–World Trade Organization protests of 1999—the so-called Battle in Seattle—took place just a couple hours away, and resonated deeply with a young Peter McCoy. Already engaged with the local food and social justice movements, he was inspired by the decentralized, guerrilla coverage of the WTO protests by Indymedia, and the countercultural discourse articulated by alternative outlets like the anarchist publishing network, CrimethInc., based in nearby Eugene, Oregon.[11] "Make media, make trouble" is its motto.

"The Battle in Seattle had just happened a few years before, and there was sort of a wave after that, that I was sort of swimming in," McCoy recalled, adding that the anarchist literature that had inspired him back then wasn't quite so compelling all these years later. "There's fanciful stories of like, the beauty of dumpster-diving and train hopping, but also just like building community and creating alternative culture, but in this really romantic way that's quite good to read when you're twenty-one. And all this was not just talked about, but it was really embodied in Olympia. So there were a lot of volunteer groups that maybe, below the surface, had sort of a political agenda, but it was really so much about building a local community, notions of local food, alternative economies were often discussed."

Born and raised in Portland, Oregon, in his early twenties McCoy ventured east for a year, where he circulated among various activist communities in New York City. Having first been turned on to fungi by his brother, who suggested growing mushrooms alongside the vegetables in their home garden, McCoy began his mycological studies in earnest. In New York, he learned that the activists he hung out with and looked up to, along with the media he consumed, generally had no concept of fungi, let alone their ecological importance.

"Every time I would read something and I would say, 'Hey, where the hell are the fungi?'" McCoy recalled. "I'd watch a whole two-hour, killer documentary about the environment, and nobody ever says a word about fungi. That stood out to me. I tried to talk to other activists, but they didn't know what I was talking about. Most of them gave me a blank stare."

Notions had begun forming in McCoy's mind, at first about what fungi could add to conversations about food systems and social justice, and what barriers stood in the way of connecting those insights to society at large. The impulse gradually gave way to a picture of fungi as both an ally and organizing principle. In 2007 he returned from New York and enrolled in Evergreen State with "all these mycelial metaphors stirring in my mind." It proved at first to be a solitary preoccupation. But before long, not unlike a pair of spores carried into the same coffee shop by a gust of wind, he encountered someone who understood.

"I was wearing a mushroomy T-shirt and serendipitously sat across from this person, Maya, who was the first person really that was not only interested in mycology but also super involved with environmental issues," McCoy recalled. Maya Elson and he quickly struck up a friendship, and together worked out how to unpack and realize the messages in the mycelium.

After gathering a handful of their fellow local mycophiles, they founded the Olympia Mycelial Network. The OMN provided free workshops to teach mushroom-growing skills, constructed mushroom gardens in and around the city, and generally advocated for the remedial and medicinal benefits of fungi. Its members soon went on to start other projects, such as the local women-run mushroom farm, Myco-Uprrhizal, as McCoy, Elson, and company continued incubating concepts for a new kind of socially engaged mycology. At that time, the world of popular mycology was already being reshaped by a nearby myco-evangelist whose star was beginning to rise.

"Paul Stamets was just starting to blow up," Elson told me. "We were super inspired and super intrigued by his work, and really excited about it, but we really wanted there to be a larger grassroots DIY movement, and we wanted to have free workshops, and we wanted to make it not about any kind of business or commercial gain. And we wanted to tie it into a deeper social justice and biocentrism theme. So we did lots of workshops, and as I started to explore this field of applied mycology, I realized there were different people in different places doing stuff, but they weren't necessarily talking to one another."

After another two-year stint in New York, McCoy returned once more, still inspired by the subversive potential of independent media and radical political movements. He spent the next six months organizing the tangle of mycosocial ideas into a zine, published in 2009 under the auspices of the Spore Liberation Front. "We, the members of the Spore Liberation Front, see the lifecycle of mushrooms, and especially this mycelial stage, as a metaphor for the way humans can choose to interact in and with Gaia, our one world," it announced. "Just as mushrooms use their abilities to share nutrients with plants and break down toxic chemicals to keep their microcosm cleaner and healthier, so can we as humans live committed to the health of our planet through our natural role as stewards and caretakers of the land."[12]

The zine's thirty-five black-and-white pages could be described as information-dense, a lovingly illustrated guide to mushroom identification, cultivation, ethnomycology, and applied mycology concepts. Part manifesto, part citizen mycologist primer, its informational illustrations of gill types and reproductive cycles shared space with short treatises on the philosophical and political implications of mycology and the fungal metaphor. Learn the life cycle of fungi, for instance, then consider the way this same cycle describes the process of people, with like-minded activists meeting and forming a network that eventually takes the solid, insistent form of a mushroom, which then melts away as the spores disperse and the cycle repeats. Little wonder the zine starts out with "A Call to Sporulate," and that's just what it did, spreading more widely and quickly than anyone involved in its creation expected.

"I thought that nobody else would read it, because I had met nobody else that cared about these intersections," recalled McCoy. "I thought it was just for me and Maya, but then eventually we started to go to zine distros and events and book fairs, and then eventually sent them to zine distros all around the world. And soon enough I had sold or distributed a couple thousand of them. It was like, 'Oh wow, maybe people actually care about this.' And then eventually Maya was the person who suggested we do the first convergence."

As Elson recalled, "I had something between a dream and a vision of all these mycelial networks converging in one spot, and creating a giant mushroom of an event."

Two years after the publication of the zine, early in September of 2011, the first Radical Mycology Convergence took place in the town of Concrete, Washington. It was a free, volunteer-run gathering, focused on sharing mycological knowledge and skills, with the intent to "bring together people of all backgrounds and abilities to destigmatize and simplify this information through the engagement of various learning modes while fostering a network of like-minded people."[13] Remediation of damaged environments featured high on the agenda, alongside low-tech, low- or no-cost techniques for growing culinary and medicinal mushrooms, with the hope of building a widespread, decentralized network of mycological interest and engagement, emphasizing ecological and human health, nondiscrimination and accessibility. The idea resonated.

"When we first started talking about it, we were like, oh yeah, it'll just be like fifty punk kids hanging out in the woods talking about mushrooms. We'll talk about growing and remediation, it'll be rad, and we'll have a potluck or something," McCoy told me. "It wasn't until we did the first convergence that I really realized that this was something of great value, that was really missing, and I decided from then on to put most of my time and energy into it."

The weekend-long event for many marked a revelatory moment of "finding the others," as psychedelic evangelist Timothy Leary entreated his followers to do. After the second convergence in Port Townsend, Washington, in 2012 McCoy, Elson, Willoughby Arevalo, and Mara Penfil came together as members of the semiformal Radical Mycology Collective. The convergence became a biennial event, and in 2014 the collective set out on a forty-stop tour of the United States, including the next RMC in Illinois, on land stewarded by "radical ecologist" Nance Klehm. From their "mobile mushroom lab"—basically an RV stuffed with mushroom-cultivation equipment—they traveled the country, facilitating workshops and extending the proverbial mycelial network. "It was exciting to be enmeshed in this culture of exchange," said Arevalo. "And it was so modeled after mycorrhizae, for example, and ecology in general."

The profile and network of Radical Mycology were growing rapidly. After the 2016 convergence in upstate New York, however, mounting interpersonal stresses came to a head, ultimately dissolving the collective, and its core members went their separate ways. A unifying sentiment among all involved though is that the convergences marked uniquely powerful moments of community building, the principles and intentions of which remain vital in efforts toward working with fungi to achieve a regenerative, reciprocal relationship with the planet—similar principles to those found in permaculture, or a "biocentric" view, as Elson describes it.

"I do feel like we started at a really crucial time when this kind of stuff was just starting to pick up momentum," she said in retrospect. "All of the convergences were super juicy, but that first one had this special energy to it, because it was the first time a lot of these people had ever connected with other mycologists who were into the same kind of stuff that they were. People were growing commercially, sure, but these were people who were into the fungi because they love the fungi, and they want to work with fungi for healing the earth."

———————

Mycelial metaphors are apparently inescapable in the realm of Radical and DIY Mycology.[14] I chuckled aloud upon first encountering unironic use of the word *insporation*. Yet the more one learns about fungi, the more the comparisons make sense.

The fungal frame of reference can offer an intuitive interpretation of various situations. Viewing an environment—social, geographical, personal—as a "substrate," with its landscape of "nutrients," "symbionts," and "competing organisms" can render maps of goals, obstacles, and opportunities in greater relief. For some, they offer no less than a model of how to structure society, build an education system, or mediate inter-personal conflicts.

"It was super fringe before," said Jason Scott, one of the 2018 con-vergence organizers and proprietor of a medicinal mushroom tincture company called Feral Fungi. Scott represented some of the unlikely alliances that can form between mycology and other disciplines. In his mushroom extractions, he had found a practical way of bringing together his interests in fungi and the also-esoteric realm of alchemy. Indeed, among the seemingly endless range of mycological metaphors, fungi are often referred to as "nature's alchemists," because of their remarkable enzymatic acumen.[15]

"I found myself in this really weird niche where I'm into alchemy, and I'm into mushrooms, and both are pretty fringe topics," Scott said.

"Mushroom people are typically pretty weird, and alchemy people are pretty open-minded, but that's what they have in common: they're fringe topics, and the culture is wide, and they both offer this deeper understanding of the natural world and of nature, and how to live in more ecological balance."

Laura Kennedy, another of the three organizers of the 2018 RMC, stood at the edge of the festival reviewing her clipboard as events got under way, tiny polypore mushroom earrings jostling with each turn of her head. Kennedy and her partners at Rise Up Remedies, a small organic medicinal herb and mushroom farm just outside Portland, had been working to gradually bring their neighbors into engagement with fungi for their capacity to enrich soil, recycle agricultural waste, and generate extra revenue.

"There are fourteen other farms on the same land," she told me. "They're doing so much in vegetable production that they haven't really considered the possibilities of using mushrooms. So we're bringing the mushrooms to this farm space, where there's so many farmers producing food, and really setting an example for another way of doing it, another product that adds a lot of value, in terms of selling it but also ecologically. So the analogy of the mycelia kind of reminds me that we're like the little hyphal tip that's just coming in and starting to make our way."

Fungi directly informed planning and decision-making processes at the RMC, down to questions of where to place porta-toilets; how might a fungus distribute these resources? But these fungal heuristics are also aimed at creating a shared and accessible vernacular, part of an effort at bringing together folks from different backgrounds and levels of expertise. The mycelial metaphors helped to drive home the ways that fungi themselves operated, complemented by a practical set of skills for mushroom cultivation in which nonscientists could easily participate and bring unique perspectives.

"At a certain point you have to overcome a language barrier, and create new language to describe what we're getting at, because language is often used as a hurdle to understanding," McCoy said. "I'm all for science, I want to make sure we're not helping bad information spread. But I'm also interested in just, like, going out there and saying crazy, weird, interesting stuff that's never been said, and thinking philosophically or in abstract ways. Just saying what if? That's my favorite line with fungi, just, what if?"

The notion of spreading spores took clearest form in the context of education; by equipping people with knowledge about fungal biology, ecology, and the potential applications of fungi, it created a "social substrate" for the "myceliation" of the community as it spread. Once "inoculated," we would each take some part of the experience home with us, to find purchase in the substrate of our own lives and communities.

Metaphors, though, sometimes cut across purposes. "I think the most important part is to remember that fungi represent the good and the bad, and to use only positive language around them kind of takes away their agency," said Mara Penfil, noting some of the troublesome aspects of the mycelial metaphor. "It's literally this white mass that's colonizing brown matter."

Founder of the feminist mycological collective Female and Fungi, Penfil also helped organize earlier convergences. Together with my friend, farmer and amateur mycologist Alanna Burns, and soil biologist and author Leila Darwish, they had begun forming their own node in the North American "mycelial network." Though connected with Radical Mycology and sharing much of the same language and vision, Female and Fungi was not a franchise; in fact, it was birthed at Telluride Mushroom Festival. "Me and my friend were walking around, noticing the male dominance of mycology," Penfil told me. "There were many females there, but only three were presenting, and one wasn't even talking about mushrooms, they were talking about spiritual woo-woo stuff."

There are numerous other groups that could be regarded as within the same movement that ethnographer Dr. Joanna Steinhardt described as DIY mycology. It could include DIY Fungi, Danielle Stevenson's company in Vancouver; also, Bay Area Applied Mycology in California; Fungi for the People in Oregon; the Central Texas MycoAlliance; even Smugtown and the Mycelium Underground, although there is a certain East Coast–West Coast divide even within this often tight-knit mycelial network, and these terms don't describe clubs or group as much as communities and a certain set of practices and attitudes. Despite its name, the convergence wasn't really the center of anything in particular, and that was just as nature showed it should be. Diversity, after all, makes for a healthy ecosystem.[16] Taking our cue from the metaphor, this ought to apply to society, too, but one hopes we don't need mushrooms to tell us that.

Largely unspoken at the RMC was a vision of interconnectedness that is often associated with the psychedelic experience, which was the subject of a couple of talks, but compared to Telluride, seemed a relatively low priority on the agenda. It was officially a drug-free event, although substances were treated with a certain laissez-faire attitude so long as people were discreet and conducted any funny business downwind and well away from children. Of far greater concern, as indicated by its prominence in the event literature, was abusive or oppressive behavior.

"A big part of this is teaching, but also letting down our guards," McCoy told me. "At the end of the day, a lot of people aren't going to be interested in learning mycology, but a lot of people, if you don't make them feel bad for not knowing about it, they might become curious. Especially nowadays, when fungi are increasingly popular."

The increased popularity comes with growing concerns about the possible repercussions, and concerns about equity and enfranchisement. *Psilocybe* mushrooms are known to have held sacred status in South American, Central American, and in other Indigenous cultures for centuries, even millennia. Yet as psychedelic therapy quickly grows into mainstream acceptance, its seems poised to deliver certain people— mostly white—mass profits, even as many others—disproportionately people of color—face prison for their use and distribution.[17] Companies such as Compass Pathways—which has successfully secured proprietary means for large-scale manufacture of psilocybin—have become controversial for what seem to many an effort to capitalize on a massive emerging market, creating tension between the pursuit of scale and the honoring of the sacred.[18] Thus *Psilocybe* decriminalization has taken its place squarely at the intersection between fungi and social justice.

One of the panels at the RMC was facilitated by artist and activist Paula Graciela Kahn. "I want to take this time for us to all close our eyes and to honor the original peoples of this land," she began. Kahn's crowded Saturday morning talk, titled "Justice for Psilocybina," focused on the commodification, co-optation, and globalization of *Psilocybe* mushrooms and other sacred plant medicines. This was also the subject of a workshop that Kahn organized later that same evening, inviting attendees to participate in group discussions about how to approach a strategy for broader decriminalization.

As the sun set with a luminous flourish behind the trees, I joined a sitting circle of about two dozen people that had formed around Kahn near the big red barnhouse. She asked everyone to weigh in on the criminalization of psilocybin, why it mattered, and what was at stake. We were then instructed to form small breakout groups based on our particular interests or areas of expertise; naturally, I joined the media strategy group, which, perhaps unsurprisingly, consisted entirely of fellow white folks, skewing toward middle age. Regrettably, the discussion in our group struck me as rather shallow, preoccupied with internet memes and using social media to "raise awareness." While such things are certainly useful—even essential—to any effort at public engagement in the twenty-first century, the lived experiences of injustice at the heart of the issue didn't seem to resonate.

I wasn't the only one who noticed. As the full circle reconvened, each breakout group was invited to share our "findings." I had been designated as the media group's spokesperson, and frankly felt reluctant to recount our conversation which had, in my opinion, generated little insight. Just before my turn came up, a young person who identified themselves as Natalie, stepped forward to interrupt the proceedings. "You guys need to think of your privilege," they said with some evident effort at maintaining a calm tone. "There's so much trauma that we have from colonization, from capitalism, all these things. It's very sacred medicine, and it was kept secret for a reason." Natalie disclosed that they had made the trip to the RMC from Oaxaca, Mexico, the place from which banker R. Gordon Wasson first brought back tales of magic mushrooms.

Kahn received the critique gracefully, and offered an account of her own work in centering Indigenous communities as reassurance that she understood what Natalie was saying, even if it wasn't reflected in the discourse beside the barn that night. The issue wasn't resolved there, of course—how could it be—but the two sides seemed to have heard each other, and it occurred to me that the tense exchange wasn't an aberration; perhaps it was the very value of a gathering such as this. Extending an open invitation for engagement means reckoning with sometimes sharply distinct experiences and perspectives. It would have been more telling if no such tensions emerged at a gathering ostensibly aimed at diversifying participation and elevating social consciousness. In gathering to better understand and work with fungi, perhaps, there is also opportunity to improve how we work with and understand one another.

Myceliated State

The summer after my visit to the Radical Mycology Convergence, the New Moon Mycology Summit gathered near the town of Thurman, in the southern Adirondacks of New York. Driving into the area around midnight, I parked by a hand-painted sign on the side of a pitch-black road, crossing to the visitors' table, where a pair of volunteers wearing headlamps waited under a canopy to register late-comers like myself. Before I could get my cloth scrap wristband, though, I had to sign a waiver.

"Language that is discriminatory, derogatory, or prejudice[d] towards a specific race, religion, and/or the lgbtq plus community is unacceptable and will not be tolerated," it read. "New Moon organizers, coordinators, staff, and volunteers will not tolerate inappropriate behavior involving language, and/or touch that is non-consensual. Physical or emotional threats and harassment are not allowed. Period."

The terms were clear, and I happily signed, trudging into the dark with tent and sleeping bag in tow. Negotiating dirt roads and dense corridors of trees, I passed a sleeping farmhouse as I worked my way toward the glint of a bonfire, stopping to warm up and chat with a few of the folks still yet to call it a night. Then I pitched my tent and tucked in under a cloudless, glittering sky.

In the morning, I awoke to a strange sound coming from somewhere downhill. As I walked toward the sound, it resolved into a quiet, nearly whispered chant in steady eighth notes. "Fungus, fungus, fungus, fungus . . ." In front of the farmhouse from the night before, I discovered the source: a circle of some fifty people were softly singing a camp song, of sorts.

"If you touch a milky mushroom you will find it oozes goo / If you eat an amanita that will be the end of you," sang the group, led by the handful who knew the verses until everyone joined in for the easy-to-remember chorus:

Oh mu-*shroom madness!*
They're growing all around
Some like to grow on old dead trees
And some grow on the ground—Hey!

Fungus, fungus, fungus, fungus
If you step into a forest and it's glowing in the night
You can bet it's fungal hyphae bioluminescent light

Oh mu-*shroom madness!*
They're growing all around
Some like to grow on old dead trees
And some grow on the ground—Hey!

The sing-along was part of the first morning's check-in ritual. A ring of yawning, stretching mycophiles sipped their coffee and reishi tea and listened as organizers acknowledged the traditional Mohawk and Haudenosaunee land on which we had gathered. Then the group went over introductions, intentions, and logistical considerations, offering a glimpse at the overall composition and character of this group, which was notably more diverse than at any other mycological event I'd yet attended.

If the energy and attitude of the gathering weren't already clear, the contents of its schedule made the point. "Exploring Mushrooms and More-Than-Human Personhood"; "Underground Struggles in Defense of Earth"; "Decomposing the Toxic Narrative"; "Mycology Is a Queer Science." Cultivation, cooking, foraging; a smattering of vendor tents, all seemed attendant to the conversations being had about decolonizing foodways and medicine; de-patriarchalizing (or *rematriating*) mycology and science generally; decriminalizing and reclaiming ancestral medicines; challenging gender and other binaries. I had never heard the concept of "deadnaming" a mushroom before visiting the NMMS.[19] Nor had I ever seen a bag of American Spirit tobacco take on a sense of the sacred, but that's just what happened when Mario Ceballos of the POC Fungi Community led the morning circle in a ceremony. Later, he would lead a panel discussion on the decriminalization of *Psilocybe* and other traditional plant medicines.

Driving the agenda was no less a mission than dissolving the obstinate stumps of heteropatriarchy, racism, and extractive capitalism, with a commitment to center the land, its original inhabitants, and their traditional ways of knowing. Composed of people from around the country, here was a "mycelial network" guided by clearly defined principles and a clear intention to persist and to grow. Wandering the grounds after breakfast, I realized that in the darkness of the previous night I had walked right past a large hand-painted banner strung up between two trees. It read: MYCELIATE THE STATE.

———

The New Moon Mycology Summit was organized primarily by Olga Tzogas, Nina O'Malley, Charlie Aller, and Doğa Tekin, together forming the informal core of the Mycelium Underground. The name, I learned, deliberately evoked a number of images: the Underground Railroad, the Weather Underground, and of course fungi themselves. In the latter sense at least, the summit lived up to its organizers' collective moniker. A hidden network had given rise to an ephemeral but tangible expression meant to reify and propagate its essence, and cast proverbial spores. I was starting to get the hang of these mycelial metaphors.

By the time I reached the NMMS, it was already clear that something resembling a movement has been building around fungi. What exactly to call it, though, was elusive. If it was a *movement*, it was—appropriately enough, given the fungal theme—multifaceted and intersectional, led by its outermost edges. "DIY mycology is not quite a full-fledged movement, although it seems to be taking on the outlines of one," writes Dr. Joanna Steinhardt. "As such, practitioners are not united under one identifiable name. Some call themselves 'radical mycologists' or 'applied mycologists,' but most are unbothered by the question of what their pastime might be called by those outside their social circles."[20]

If we are to take the mycological metaphor seriously, we can't consider the Mycelium Underground to be about one, four, or any other discrete number of individuals. "We're sort of using that moniker to describe ourselves," said Aller. "But I hope it becomes something broader than that. It really wasn't ever just the four of us."

"We wanted it to be, not amorphous, *anonymous*," said O'Malley. "Well, amorphous *and* anonymous!"

This second NMMS brought together a group of about three hundred people. Since the last gathering, the themes of which had centered largely on oil and gas and direct action activism, things had grown more organized and focused. The all-inclusive ticket was available on a sliding scale price of $100 to $450, but no one would be turned away for paying less, nor for an inability to pay. Land acknowledgments were all but ubiquitous, stated recognitions of the Indigenous people whose violent dislocation had cleared the way for our fungal frolic (and indeed all of modern American society); during a lunchtime panel, representatives of the Northeast Farmers of Color Land Trust presented a walkthrough of how land factored into redistributive justice. It was normal and encouraged to ask for pronouns. Most locative names were pointedly preceded by the word *so-called*; we were in so-called Thurman, New York, but really we stood on Mohawk and Haudenosaunee land. LGBTQ and BIPOC were centered in the proceedings and women were predominantly in roles of leadership.

For all the consciousness about issues of social, historical, and ecological inequity or injustice, a certain lightheartedness also prevailed; Tess Burzynski led a course on the so-called "bleeding tooth" fungi, in the genus *Hydnellum*, helping her audience remember the taxonomic ranks of domain, kingdom, phylum, class, order, family, genus, and species with the mnemonic: Don't Kick Puffballs 'Cause Other Fungi Get Suspicious. Rikki Longino and Oona Goodman, who biked across several state lines with their herbal tea and tarot booth in tow, presented chocolate truffles to customers with a warbling wail that started and ended with the lifting and closing of the lid of the serving platter, as if the ghosts of opera singers were trapped inside. Willoughby Arevalo came all the way from Vancouver, British Columbia; in a garage on the edge of the property, he laid out a large "smell wheel," an entertaining effort at mapping the notoriously subjective dimension of odor in identifying mushrooms, which gradually piled up atop sections labeled with a greater variety of descriptive terms for olfaction than I knew existed: chemical, sweet/spice, nutty, yeasty, resinous, turpentine, fabaceous (beany).[21]

One night, Arevalo delivered a performance of *The Sex Life of Mushrooms*. "Bring yourself into the world of a hypha," he said in a tinny, deliberately stilted cadence from behind a microphone with a polypore mushroom stuck to it, recounting the fungal life cycle in suggestive entendre that drew giggles from the crowd. He went on to comment on fungal modes of reproduction, such as that of puffballs, a "gasteroid" fungus, so-called for their storage of spores as an internal mass. The comparison didn't sit right with him. "Do you know anybody who makes their reproductive cells, their gametes, in their stomach?" Arevalo asked the crowd, slowly but incredulously, before snapping back into a conversational tone and volume. "*Do ya!?* It was probably a man that came up with this term, *gasteroid*. If it were a woman, maybe she would have called these fungi the *uteroid* fungi. There's a lot more similarity."

The proceedings were leavened with live music, beers around the bonfire, foraging, skill shares, and free-associative conversations with new friends at mealtime—courtesy of Seeds of Peace, an organization that supports activists undertaking nonviolent direct action around the United States. Conversations started easily among strangers of various backgrounds and identities, sharing not only an interest in mushrooms but also the hope for a world that better reflected the diversity and values on display in Thurman that weekend.

"Gatherings are potent," said O'Malley. "It's powerful! There are a lot of people's energies going towards different things, and we wanted them to be going towards what we thought was degrading the structures that no longer served humanity as a whole, while growing lifeways that better served it."

Even the most modest shifts toward equity, such as decentering white cis men, can make for a dramatic difference in any space. Probably, that's because any such space occurs in the context of a society that historically enforces against that. Around mushrooms, there was an unusual opportunity to organize people from a variety of backgrounds and identities to share ecological concepts useful for land stewardship, to teach cultivation skills and connect them to issues of food and medicinal sovereignty, to examine citizen science as a way of empowering marginalized communities with tools to better understand and engage with their environments. Mushrooms and fungi had gone from being the subject at hand to the platform on which broader conversations could be had. Anyone attending with an interest solely in picking and learning about mushrooms would find plenty to occupy their time and attention. But they would have to put effort into avoiding conversations about environmental racism and climate change, equity and justice for LGBTQ and BIPOC communities, the urgent need to supplant patriarchy and white supremacy. At a certain point, calling it a fungi festival seemed like missing the forest for the mushrooms.

Truly looking to fungi, just like looking for them, means looking through them: to the trees, soil, water, and air, and indeed to the broader environmental and social realities to which all of the above are also connected. Talking about fungi necessarily means talking about trees and insects and soil, which means talking about forests and landscapes, which means talking about oil pipelines and capitalism and colonialism. Maybe working with fungi, or organizing around them, also offered a means of realizing change at the levels of community and society.

"Landscapes are not backdrops for historical action: they are themselves active," writes Anna Tsing in her splendid book *The Mushroom at the End of the World*. In examining the culture, ecologies, and economies surrounding the matsutake mushroom, and the role it plays in the collective organism—or perhaps, society—of the forest, she makes another kind of land acknowledgment. "Matsutake and pine don't just grow in forests; they make forests. Matsutake forests are gatherings that build and transform landscapes."[22] A spore lands near a tree, and eventually the forest changes through an expanding set of mycorrhizal relationships, but neither tree nor fungus are singularly responsible. That's how life works, and maybe it's suggestive of how society can work, too.

As Tsing points out, fungi seem prone to undermining the human systems and logics that carve up and capitalize upon an interconnected living world. Wood rot fungus, dread of homeowners the world over—in particular in Europe and the U.K.—was mostly localized around the Himalayas until the nineteenth century, when the British Empire entered the region and returned with warships inoculated by the house-munching mold.[23] Are fungi therefore anti-colonial? The potato blight of 1845 affected not just Ireland but all of Europe, Tsing observes, however it devastated Ireland's potatoes in particular—and therefore Ireland as a whole—because of the reliance on that single crop; fungi, like nature, apparently abhor a monoculture. Are fungi anti-capitalist? Not exactly, at least I don't think so; such claims are enticing, but reductive. But perhaps fungi do demonstrate that the principles of extraction and the enforcement of arbitrary borders are not the recipe for a successful ecosystem or society.

On the last night of the NMMS, I caught a lecture by Tiokasin Ghosthorse, a member of Cheyenne River Lakota Nation of South Dakota and popular speaker and radio host. To an audience that sat in rapt silence, he mused on the importance of understanding one's own language. Crickets sounding in the pauses between his words, he contrasted the nounless, integrative language of Lakota with the parochial, possessive structures of English syntax and grammar. In the process, he pointed to the absurdity and humor inherent in words like *technology* (tech-no-logic), and *America*, in which he saw the true meaning as "love for riches."

"Mycology," Ghosthorse said, then paused for a moment as the word resonated. "I'm not going to talk about mycology. Mycology is already talked about. That's why you're here. You complete the thought here."

CHAPTER EIGHT

FeRMeNT YourseLf

The landscape around Short Mountain, curiously the tallest point in middle Tennessee, is a tangle of creek-cut ravines crowded by tulip poplars and thick oak-hickory forest. The local dogs have a funny habit of pacing alongside the passing cars on the serpentine dirt roads and napping in the shaded embankments. My cell phone reception vanished along with the pavement, well before I crossed the leaf-littered gulley and ascended a steep, rutted road that led to the home of Sandor Katz. Near the top of the hill, a hand-painted sign strung to a tree welcomed me to Walnut Ridge.

I had come to the deep reaches of Cannon County to attend a semiannual five-day workshop held at the off-grid homestead of the celebrated fermentation revivalist. Parking next to a half court–sized solar panel, I approached the two-hundred-year-old, handsomely refurbished cabin. It was still morning, and quiet save for the murmur of Katz's distinctive voice drifting from behind a side door, and the complaint of a small cat that stalked out from the trees in hopes of getting inside. Creaking open the door, I stepped into the midst of a crowd clustered around a large countertop at the center of a green-walled wizard's lair of a kitchen. "You must be Doug. Welcome, welcome," Katz said, easing my slight embarrassment over the interruption. I tucked up against the stone backside of a huge fireplace as he resumed his demonstration.

Katz stood at the head of the gaggle of more than a dozen aspirant microbe-charmers, eager to plumb the seemingly infinite culinary possibilities of fermentation. Sagacious and scruffy, with a salt-and-pepper beard and a perforated old T-shirt that read PLEASE PASS THE BACTERIA, he massaged a heaping teaspoon of koji starter into barley, gingerly dumping the mix into a cedar tray box and furrowing the pile with his fingers, explaining that he was increasing the surface area to radiate heat and avoid killing the fungus while it incubated.

Fermentation might be most easily associated with bacteria, but fungi are no less fundamental to fermentative practices. Koji is a prime example. The *Aspergillus oryzae* mold is fundamental to miso, soy sauce, amazake, sake, and other traditional ferments; in Japan it is the "national mold," and its use in China extends back some nine thousand years.[1] But koji has undergone a renaissance in recent years, as chefs and gastronomists and home fermenters alike discover its many applications for bringing umami out of all manner of ingredients. The malleability, reliability, and relatively easy application of koji have made it an intense subject of interest for those looking to unlock new flavors through fermentation, such as for meat and vegetable charcuterie.

"Koji's all about the enzymes," said Katz, "because koji's not really a food, people don't really eat koji, although it's delicious, and there's nothing wrong with eating it, but generally, it's a *vehicle*."

Setting the barley into his incubator—an old refrigerator with a tricky door, stripped of its wiring and retrofitted with an incandescent light bulb—Katz retrieved a small crock of another koji product, called doubanjiang. A burgundy paste made with chili peppers, fava beans, and soybeans, he'd brought it back from the Sichuan region of southeast China, where he and fermentation entrepreneur Mara Jane King had documented the country's vast landscape of fermentation traditions.[2] A teaspoon's worth touched my tongue, instantly igniting a fireworks of flavors: complex, spicy, smoky, tangy, and totally novel, at least to me.

As the wave of pungent sensation slowly ebbed, I noticed that splashes of red, blue, and yellow light had settled on the cluttered kitchen counter. Looking up, I saw the source: a stained glass window wrought in the shapes of bacteria. In Katz's practice, age-old traditions were approached with casual, accessible immediacy, undertaken with an "anyone can do it" attitude that didn't see partaking in tradition as requiring a biology degree, nor as exclusive to those who grew up with grandparents who still practiced the old recipes.

For some twenty-five years, Katz has been an ambassador of and advocate for fermentation; the mustachioed maestro of all things bubble and brine. He is also a longtime champion of local, resilient, diversified food systems. The author of *The Art of Fermentation* and other influential books, he has traveled the globe learning, adopting, and teaching about the seemingly endless varieties of rich fermentative practices refracted throughout the world's cultures. Sharing these traditions was less about conserving them, though, than inspiring others to carry them forward in their own contexts.

Jeff Gordinier of the *New York Times* found an apt description for Katz, saying that he "has become for fermentation what Timothy Leary was for psychedelic drugs: a charismatic, consciousness-raising thinker and advocate who wants people to see the world in a new way."[3] Meanwhile, and thanks in no small way to his influence, fermentation has been undergoing a renaissance in North America and around the world. From apartment windowsills and basement incubators to the test kitchens of double Michelin restaurants, enthusiasm is frothing over for the flavors, potential health benefits, and mystique of transforming food with microbes.

Marketing firm Pollock Communications' eighth annual "What's Trending in Nutrition" survey of some registered dietitians named fermented foods as the "number one superfood" for 2020, the second year in a row.[4] In 2018, consumption of fermented foods more than doubled over the course of a year.[5] Globally, analysts expect the market for fermented foods to reach nearly $700 billion by 2023.[6] In haute cuisine, too, fermentation is ascendant. According to Didier Carcano, DuPont Nutrition and Health's global marketing and strategy leader for cultures and dairy probiotics (how's that for a job title?), fermentation is a bona fide "mega-trend."[7] New York–based gustatory celebrity David Chang dedicated a lab to developing new fermented products, while across the Atlantic in Copenhagen, René Redzepi's famous Noma restaurant had taken to including a fermented item with almost every plate.[8] "Fermentation isn't responsible for one specific taste at

Noma," wrote Redzepi and David Zilber in their hit book *The Noma Guide to Fermentation*. "It's responsible for improving everything."[9]

Earning the marketable moniker of *superfood* is a sure sign of trendiness. But a trend also tends to denote something new, and there is nothing new about fermentation. The practice of making mead from honey stretches back over ten thousand years, suggesting that fermentation is at least as old as agriculture.[10] The movements of people through time and space can be traced in fermentation traditions. Sauerkraut—the basis of Katz's nickname, Sandorkraut—might call to mind the cultures of central Europe, but consensus has it that the sharp, crisp fermented cabbage actually originated in China more than two thousand years ago.[11] The myriad roles of microbes—not only bacteria but yeasts and other fungi, as well—in preserving and enhancing food are not even a uniquely human concern. Tree shrews, for example, and slow lorises eat the fermented nectar of flower buds. Moose are known to get drunk off apples fermenting on the ground.[12] The key difference, of course, is that human fermenters have established intention, ritual, and tradition around the practice of invoking bacteria and fungi to improve food and drink.

It's also not as though fermentation ever went out of style; as is often pointed out, rare is the refrigerator or pantry that doesn't contain at least one fermented item. Anyone who drinks coffee, eats bread, or drinks beer has caught the fermentation "wave." Merely pointing to sales of kimchi or kombucha fails to tell the story of fermentation's foment. Indeed, the widening adoption of fermented food and drink can't be pinned to any one factor, but it is concurrent with a general move away from processed, nutritionally empty foods and toward an embrace of locality, diversity of gut-benefiting microbes, and craft. There is real power in the idea that you—yes, you—can wield bacteria and fungi to enhance the nutritional and culinary and thrift value of your food. It might sound dangerous, or highly specialized, and sure, cheese makers and wine makers and sake brewers are often experts with deep and special knowledge of their tradition, and there are potential risks to health should the wrong bacteria take hold of a ferment and someone eat it. Yet those traditions were begun and carried through time by generations of people who had never heard of a microbe; they learned through experience and observation and the passing of knowledge from one generation to the next through oral tradition, demonstration and participation, and in the modern era, increasingly through books and other media. The practices are often quite simple, because they had to be, in order to propagate. Even if you aren't quite skilled enough to get a job as a master brewer, assuming you can take care of a houseplant and live with understanding housemates, you can almost certainly take care of a kombucha SCOBY or a wine mother.

What Katz conveys to his growing audience is about more than a new range of products or flavors; there is a philosophy implicit in the discipline and practice of enhancing, preserving, or even detoxifying food, reducing waste and stretching value, deepening our relationship to what we eat and to the microbial communities in our guts, comprised of fungi and bacteria alike. That philosophy isn't always explicitly articulated; whatever the path or reason that leads someone into his kitchen, Katz is seemingly glad just that they are interested.[13] He's a revivalist, not a purist.

"My most important goal is making information available for people who want it," Katz told me. "I'm not invested in whether people want to make their own. Like, if people would rather go and buy sauerkraut that someone else is making, I don't feel like that's a bad thing. But when I hear people whine about how expensive it is, like, 'Oh, I don't want to pay eight dollars for a pint of sauerkraut,' then I'm like, well, yeah, you could buy those vegetables for a dollar and chop them up yourself and salt them, and I've been trying to tell you how easy that is."

Throughout the workshops, the days took on a distinct rhythm. After a few hours of demonstrating a basic ferment—tempeh, miso, yogurt— or preparing something for one of the day's meals—such as dosas or buckwheat bread—the morning's projects were set out to dry or slid into the incubator. Inevitably, a crock would emerge from some hidden recess, filled with sauerkraut or nuka pickles or another delicious morsel teeming with microbes. It occurred to me how often the functional life of the food I usually ate ended before I even brought it home.

By the time lunch or dinner rolled around, the broadest suggestion of a meal plan would catalyze a commotion as everyone found a role in the kitchen. The depth of culinary expertise in the room was profound, so my contribution tended to be washing the seemingly endless stream of dishes. The room, mostly quiet throughout the day except for Katz's demonstrations, erupted into conversation and emergent collaboration when the time came to cook and eat, a chaotic concert of clanging bowls, jars, and metal cookware that culminated in a table covered with an embarrassment of culinary riches: żurek soup; eggplants grilled in sorghum miso and fermented shiitake paste; krauted onions, roasted beets, and vegan cashew cheese. (As a vegetarian, I opted out of the pickled beef tongue.) As the table filled up, a small gong quieted the excited group. "Sometimes it means 'time to eat'; sometimes it means 'break is over, come back, come back, wherever you are!'" said Katz.

Each of the students at Katz's country home had brought their own reason for pursuing fermentation, and an evident passion for the subject. Many were particularly excited by the new possibilities represented by koji. One, Tyler Grayko, worked at a restaurant in Chattanooga, where he hoped it would help to bring out new dimensions of flavors from plant-based dishes. "Koji really is the next frontier," he enthused. Another was Eleana Hsu. Among the treats she'd brought from San Francisco were koji-cultured sweet potatoes and carrots, a delicious experiment in vegetable charcuterie; Hsu had decided to quit her commercial real estate job to start a business with her partner called Shared Cultures.

"I caught the foraging bug last year because I got severely depressed," she told me. "I found, well . . . the mushrooms found *me*, and from there it was like this new world, and then learning how to preserve all of it. And then it kind of just came back to koji, because koji is a fungus. I couldn't ignore this thing in my soul; I have worked so hard to be where I am professionally in the corporate world, climbing this ladder, but it doesn't mean anything at the end of the day. All this stuff really brings you back to center, to what it means to be a human being."

Another student, Sean Roy Parker, saw the makings of social change reflected in fermentation. Wearing a shirt of his own making that read FER-MENTAL HEALTH, Parker carried a notebook stuffed with assiduous notes; having crowdfunded his way to Tennessee from London, he'd promised his funders and friends a detailed account of his experience. Across the pond, his art practice took the form of running a pay-what-you-can canteen, and teaching fermentation, pickling, and other basic foods courses at sliding scale prices, something he'd managed to do, somewhat subversively, inside high-end art galleries. Parker is a self-taught cook and far from an expert in fermentation, which to him represents part of the food-decommodifying value inherent to the practice and promulgation of these skills.

"My knowledge compared to some others' here is pretty underdeveloped," he said, "but my goal is to share the basics while embracing my amateurism, and I engage with a real range of people from different backgrounds, particularly those with low income or learning disabilities. It's not expressly about being an expert and therefore it's not my capital; for me it's important more people can access the experience and take something away."

Not unlike what was taking place in the kitchen, fermentation reifies the idea that the strength of any culture is its diversity; cultural purity, like sterility, according to Katz, is a myth. Working with the latent microbes in food and floating in the air, after all, means working with communities. A kefir grain, for example, does not occur naturally; it is the result of an unbroken lineage of careful cultivation of dozens of fungi and bacteria that work on the milk into which they're dropped.[14] The kombucha SCOBY is called the mother and yet is also a daughter, so to speak, but really neither term is appropriate; each is in fact a population, a consortium; the term *SCOBY* itself is an acronym for "symbiotic culture of bacteria and yeast," which live in constant relationship to their medium of tea.[15] The fizz of fermentation is a sign of microscopic neighbors interacting. The notion of a "pure" culture—that is, specifically selected and cultivated microbial strain or strains—is a relatively recent and uniquely human concept.

In discussing yogurt, for example, Katz told the story of Élie Mechnikoff, director of Morphological Microbiology for the Pasteur Institute in Paris. In 1904, Metchnikoff accidentally inspired a yogurt fervor when in a public lecture he suggested that aging was caused by harmful intestinal bacteria, brought into the body by food. He urged boiling fruit and vegetables, to curb intake of these damaging microbes.[16] The fermented milk-heavy diets in the Balkans, where people lived unusually long lives for the time, suggested to him that the lactic acid–producing bacteria in dairy prevented the "intestinal putrefaction" that he pointed to as a cause of aging.[17] In work that would kick off the first-ever probiotic craze, and earn him a Nobel Prize, Metchnikoff identified the species that soured milk without spoiling it—namely *Lactobacillus bulgaricus*, and *Streptococcus thermophilus*—which soon became the dominant and sometimes only bacteria in mass-produced yogurts.

But try to make a starter out of one of these highly managed microbial populations, Katz observed, and you wouldn't get more than a couple of generations before the whole thing petered out. "They're mixing together two or three or four pure culture lines, and then they mix them together and make beautiful yogurt," he said as he passed his own crock of yogurt around the room.

"From that you can make a generation of good yogurt. But from each subsequent generation, it gets diluted, random environmental bacteria enter the mix and might end up displacing the original ones, or maybe just dilute the original ones, as well as these viruses called phages, which are basically diseases for bacteria, and the bacteria don't really have any defense, whereas the traditional cultures are evolved communities, like in the way that kombucha is, or some of these other things that have a structure that enables them to maintain themselves and defend themselves against random exposure to environmental bacteria." Katz called these resilient communities "heirloom" yogurt cultures.

Heirloom cultures are often maintained by way of practices evocatively referred to as "backslopping." That is, mixing a small amount of the last batch of a culture in with the next; in this way, even generations down the line, there is a molecular lineage connecting each generation of yogurt, or beer, or sourdough, or whatever the culture might be. There is beauty in the idea of an unbroken biological linkage between generations of fermentation practitioners and ferments alike, even if it might raise red flags for sterility-minded food safety regulators. As an example, Katz recounted the history of the yogurt that was making its way around the room; it started with a rag. At a workshop in the U.K. some ten years prior, the woman with whom he was copresenting had brought back an heirloom yogurt from the legendary New York knishery Yonah Schimmel. This was a yogurt culture Katz was keen to keep. But how to get it back to Tennessee from the United Kingdom? Pondering the problem, he remembered reading about a traditional method of drying cultures on a handkerchief for transportation.

"I just took a little piece of clean cloth and I saturated it with the yogurt, and I put it in the window where I was staying," he recalled. "Now I had this crusty piece of cloth. So I folded it up, I packed it in with my clothes and I brought them home, and when I unpacked, it got buried under some papers on my desk. So I was cleaning my desk a couple months later, and I come across this crusty cloth. 'Oh my God, that's the yogurt culture!' And so I made like a cup size of yogurt, and put this crusty cloth in there, tried to break off some little crusty pieces of it, and I made some beautiful yogurt out of that. And I've been perpetuating that ever since."

As demonstrated by this example, the practices of fermentation themselves travel and evolve. The culture-smeared handkerchiefs were often packed into the suitcases of immigrants, eventually finding new homes, new communities, and, over long enough time frames, new cultural expression and association. Katz noted the example of sauerkraut moving from China to western Europe and becoming a staple of the latter's food tradition. "None of these traditions are what you could describe as pure. Everything is influencing everything else," he said.

Fermentation can also trace schisms in cultural lineage. Toward the end of my visit, I asked Katz what characterizes the American fermentation culture, besides, perhaps Indigenous practices kept up in small communities. "First of all, I don't think we can just push aside the Indigenous traditions, because there are a lot of them; they're big and we don't know about all of them," Katz said. "I had a student in one workshop, a Mohawk woman, and she was really interested in traditional foods, and she said, 'I can't find any examples of fermented food in our tradition, because so many traditions got completely interrupted.' If your people got moved, you couldn't continue your food traditions, there's a lot of severing of traditions. But you know, people everywhere talk about how what's interesting about America is, all these traditions, but also no *real* tradition. People are much more open to innovation than places that are really steeped in very, very specific traditions. So maybe that's the American tradition: remixing."

In his presentations, Katz often points out the most prosaic notions of exactly what fermentation is. "Fermentation is the transformative action of microorganisms," he said to the room. "For a biologist fermentation means something different, and it's both more specific and more general.

For a biologist, fermentation means the production of energy without oxygen: anaerobic metabolism."

Of course, he noted, the koji we'd fermented with *did* require oxygen. The fungi-based, aerobic ferments are what he likes to call "oxymoronic." One oxymoronic ferment we made was tempeh, using black rice and soybeans. Thoroughly mixed with the starter culture, these were left in the incubator until the next day, when we pulled the white, fuzzy bricks out of their plastic pouches. In my hand, still uncooked, they were tangibly alive, almost sweaty.

A key difference between fungi and bacteria in food: we almost never actually see the latter. We can only detect the *evidence* of its work, mainly in texture, bubbles, flavor, and strong smells. Looking for the lactic acid and other chemical by-products is another way of detecting microbial handiwork. The word *ferment* itself is rooted in the Latin word *fervere*, "to bubble." But there is another meaning of fermentation that Katz is fond of articulating: as a metaphor, a way of understanding the process of change in arenas well beyond the crock or the incubator.

"It's about bubbling," Katz explained to me. "Like a time period where there's a lot of musical ferment; there's a lot of experimentation, there's a lot of change. People are excited and I think that excitement is the bubbliness. And I think that fermentation also can happen internally. Like when you're playing out some ideas in your mind when your mind is really active thinking about different possibilities."

By that standard, Katz's kitchen was itself a fully fledged fermentation vessel; the students gathered, and with deft guidance from the fermentation revivalist, a reaction occurred. The exchange of information, labor, and, of course, food, generated energy that needed simply to be guided, or halted momentarily by a gentle strike of the gong. In the process, community was forming, new ideas were crystalizing; students proposed ideas that Katz had never considered, demonstrating that cultural interchange is very much a nonlinear process.

On the wall at the back of the kitchen flanking the fireplace was a collection of pictures and woodcut artworks produced by Katz's fans and people in his community. At the center was a bright illustration of a single cabbage growing on a field in the mountains, with the words I

WILL FERMENT MYSELF. At first I found the phrase puzzling, but after a couple of days I was beginning to understand what it meant.

———

Central Tennessee is not the sort of place where one might intuitively expect to find a central figure in the nationwide fermentation revival. Deep red and culturally conservative, it's a place where country music and sermons play on almost every radio station. The LGBTQ community to which Katz belongs also seems an odd fit with Cannon County, where Short Mountain is located, where some seventy churches serve less than fifteen thousand people.

Short Mountain is one of more than a dozen communities around the world associated with the Radical Faeries. The movement was launched in 1979 by gay rights activist and ardent communist Harry Hay, with the goal of realizing the full potential of queer identity by creating a sanctuary in the rural core of the United States. That year, Short Mountain became home to "what is almost certainly the largest, oldest, best known and most visited planned community for lesbian, gay and transgender people in the country," according to journalist Alex Halberstadt in his *New York Times* profile of the Short Mountain Sanctuary.[18]

The piece Halberstadt wrote received a mixed response from the community it documented. To hear them tell it, residents of the Short Mountain "gayborhood" lived on good, if occasionally tense or awkward, terms with their culturally distinct neighbors. For understandable reasons, this was a community protective of its privacy.

"We always enjoyed a little bit of local obscurity and mythology, people didn't know what to make of us," said Katz. "In a way that [article] gave people more specificity to understand who we were, and maybe that's good." Among the upsides of the elevated profile, he said, "We're a lot more out now locally than we were."

The story behind Katz's own arrival to the area is well known. After being diagnosed as HIV-positive, he moved from New York City to Tennessee in 1993, diving with abandon into the practices of growing and preserving food, with a natural eye toward the potential benefits for his health. Though he is careful not to make grand claims about their health benefits, he does describe fermented foods as a critical factor in his own ongoing health. The community he found, along with the ever-expanding range of traditions and practices he learned, offered a way to live in an ever-deepening relationship with the land, and to build practical skills in defiance of what Katz once described as the problem of gay men becoming a "deskilled class."[19] He would live in a community cooperative for seventeen years before moving about a mile away to the cabin where he now teaches his workshops.

"I couldn't really see what the change I wanted was, just that what I was doing wasn't working and I had to make a big change," Katz told me during dinner after a workshop session. "I came down and visited and started thinking, maybe this is my new life. I sublet my apartment, because it would have been scary to be like, 'Okay, I'm leaving behind everything I know, and going to this big unknown,' but I was able to do it all with this sense of a trial period. But really after a couple weeks, I was like, I'm not going back to New York."

On the second day of my visit, we met a couple more members of Katz's extended community. Driving as a caravan to nearby Woodbury, the class arrived with their teacher at Short Mountain Cultures. In the kitchen of a community arts center covered in solar panels, we found ourselves enamored by a storefront dedicated to fermented food and drink.

Coolers lining the walls contained kefir, kvass, and kombucha, in glass bottles all manufactured in Memphis. Tucked into a back corner, shelves displayed ceramic crocks made in Southern Ohio; some students bemoaned that they would never be able to bring one onto their return flights. Also available were kefir grains, kombucha SCOBYs, starters for rice and barley koji, along with samples of the company's signature product: tempeh.

In a bowl were slices of tempeh that had emerged from the incubator that very morning before being tossed onto a skillet. Overnight, the mycelial threads of *Rhizopus oligosporus* had insinuated themselves throughout the bricks of soybeans; the variegated chunks of cracked legumes infused with the fuzz of filamentous fungi were savory and rich in a way that made the mass-produced varieties seem like another food altogether, a pale imitation of the real thing.

Walking into a place with kombucha on tap was almost comical for its discordance with the local environment. I couldn't help noticing the context of this small business dedicated to ferments. Outside was the very image of a small southern town; just down the street from us were a Dollar General and a Piggly Wiggly. "Doing this in Tennessee has been . . . weird," said John Parker, who started and runs SMC with his partner, Simmer. Parker's long, coarse, graying beard seemed at odds with his distinctively gentle, melodic drawl, a sign of his Alabama roots. "It was hard, you know, because people were, like, anti-vegetarian. You run into all sorts of things."

SMC began in 2016, after Parker moved to Cannon County from Brooklyn. While back East, he had received a copy of Katz's book *Wild Fermentation* and quickly became particularly obsessed with tempeh, producing it in small but constant batches out of the handmade, wooden incubator rack in the kitchen of the queer collective apartment where he lived. Parker decided to move to Cannon County after visiting for one of Katz's workshops.

Although Parker quickly integrated with the sanctuary and its extended communities, the area proved difficult for someone accustomed to the urgent social energy and professional opportunities offered by New York City. "There's no jobs, and if you look queer, you're *definitely* not going to get hired by anyone except SMC, or Short Mountain Distillery," Parker said. "It's still very conservative out here, and very religious."

After bouncing around the area for about a year, Parker decided to get back into manufacturing tempeh, this time in a spare room at Simmer's house. "I just still had the fever," he said. "My little operation at home kind of got bigger and bigger, and then I started to get a little concerned. I was like, okay, I'm not doing this with any kind of guidelines or any kind of insurance, I need to move to the next phase."

After finding an ideal location in the arts center kitchen, they got their licenses and permits in order and began making tempeh in earnest. But as soon as they started tabling at the local farmers market, they sensed a problem. "It was just like, oh my gosh, this is not going to work, because tempeh is just not very popular in Tennessee." From then on, they diversified their offerings to include kimchi, krauts, kvasses, and the rest of the fermentary possibilities.

Fermented foods, said Parker and Simmer, can be polarizing in their community, especially the more pungent varieties of fermented vegetables—people seem to either love or hate them. Krauts and kimchis aren't a standard part of the American diet, and strong smells can understandably inspire strong reactions one way or another. The difference often comes down to associations formed early in life. "Older folks that sometimes walk in the door and immediately are, like, 'That smells like sauerkraut, that smells like my grandmother's house!'" said Simmer. "We get stories about the medieval dungeon in the basement with this scary crock or the sauerkraut pounding mallet that they were just frightened of as a little kid."

In Brooklyn, one can expect a certain culinary adventurousness that is perhaps less common in middle Tennessee, although relationships to tradition often defy simple geographical distinctions. The pair noted a population of Korean War veterans in the area, who came into SMC familiar with and often fond of kimchi. But ferments like kvass or kefir aren't terms most Americans are likely to have ever even heard, let alone tasted, so Parker found a new challenge in learning to "code switch" depending on how potentially receptive a customer seemed to be. "I've made the mistake before where I'm, like, 'Do you want to try the *weirdest thing I have*, first thing?'" explained Parker. "And they're, like, 'Oh my God I have to get out of here, like, what is this place?'"

As a result of these interactions and reactions, he's learned to better read his audience, and help bring them into the fermentation fold. "It is a lot of education. Sometimes I feel like a tour director, and I've sort of developed this schtick in a way, because I know what works and what doesn't." Over the previous four years, SMC had managed to find the mix of offerings that worked for their community, which in turn had seemingly grown more receptive to fermented food and drink. The week of our visit was a record-setter for SMC, which Parker said seemed to represent a steady trend of growth. "We're just sort of amazed by it."

Despite the challenges in finding a willing customer base, Tennessee actually marks the place where the bricks of fungus-infused beans and rice were first introduced to the United States.[20] Nearby, in Summertown, is the country's largest, oldest, and most famous commune, The Farm. The Farm represents another unusual fit within the deeply traditional context of middle Tennessee. In 1971, a community of hippies drove there in buses straight from San Francisco, and set earnestly upon the project of living in accordance with a set of values that prioritized human cooperation over hierarchy.[21] At its peak, some 1,500 people lived on its 1,700 acres, each taking a vow of poverty and working in interdependent community.

To meet the substantial protein requirements of their population, The Farm turned to soybeans. Reckoning that they could get forty bushels of soybeans from an acre of land—more than a ton of beans per acre—they saw the opportunity to feed ten times as many people as they could raising and feeding dairy cows and cattle on the same land.[22] Doing this would also be consistent with their commitment to what they called the "pure vegan diet."

Among their ranks was biochemist Alexander Lyon, who helped set up a soy "dairy," which provided soy milk, tofu, and eventually tempeh. Seizing on a flurry of recent research into the cultivation of various soy products and their health benefits, they sourced tempeh starter from a USDA lab in Peoria, Illinois. A Farm member named Deborah Flowers produced two large batches of the savory bricks, and soon the Farm was culturing its own starters of *Rhizopus oligosporus*, the fungal species used in tempeh. When Cynthia Bates joined in 1974, she built a tempeh incubator out of an old refrigerator, and by the beginning of the following year, they were producing as much as two hundred pounds of tempeh a week.[23]

When the word got out about tempeh to the wider United States in 1977, The Farm became a leading source of split, hulled soybeans, and essentially became the country's first commercial supplier of tempeh starter.[24] Soon tempeh was the subject of a modest media blitz, and demand for the starters and ingredients alike spiked throughout the country. But as seems to be the way with communes, The Farm underwent seismic internal shifts, and its population quickly contracted to the current level of about two hundred people, necessitating a downsizing of the tempeh business. According to Bruce Gaynor, who took over The Farm's "Dairy" in 2018, they produce about a thousand pounds of tempeh a week for Nashville restaurants.[25]

None of this history was known to Parker, who earned the nickname John the Tempeh Man while still living in Brooklyn. When he moved to

Cannon County, it wasn't with the expectation of carrying forward the local tempeh tradition. "I ran into multiple people who spent time either homesteading at The Farm, or were children of the homesteaders, and grew up eating tempeh that they made as a community, and they were, like, 'Oh my God, this is the tempeh I used to eat when I was a kid!'" Parker said. "I'd be, like, this is so bizarre, why Tennessee? Why tempeh in Tennessee?"

I had similar questions. In the deep recesses of Cannon County, fermentation practices seemed to have taken on an essential role within and among communities seeking to forge ways of life outside the status quo. During my time there, I couldn't shake the feeling that there was something almost rebellious about the scene I was witnessing. The depth of enthusiasm and reverence for the forces at work in fermentation, and its centrality to the community, rang as profound in a way that went beyond thrift, tasty food, and intercultural exchange.

Enthusiasm could itself be considered a radical disposition, particularly in late 2019. But that wasn't quite it. Maybe, I thought, by fermenting foods, we relinquish control to forces we've been conditioned to try to impede. Rather than standing athwart the processes of biology, we could wield them, guide them, to the benefit of all organisms involved. In our daily lives, after all, we assume control over the microbial world, expecting to hold it at bay. There seemed something portentous in the collective move toward *embracing* the transformational, often mysterious activities of microbes, a matter of faith that doing so would work out to our benefit.

Here, a motley group of people had come from far and wide to establish themselves in the midst of what might be considered an unlikely, potentially even hostile environment. Over time, they had managed to gradually transform their local culinary and cultural landscape, slowly cultivating a regional taste for fermented foods, among themselves, and with the public at large, at farmers markets and on Main Street. Simply by being there and living out their lives, a transformation had occurred. The term *fermenting community* seemed it could apply here as both noun and verb, initiating transformations both inside and outside the crock.

"I don't think that there's anything inherently subversive in making yogurt or making sauerkraut; it's only in the context of a system," said Katz when I asked for his thoughts on the radical implications of fermentation. "Any of these skills, whether it's digging up a little bit of your yard to have a little garden and learning how to grow some tomatoes or some radishes, that could be a totally radical act because you're reclaiming your food. So, yes, I mean, I think it can be subversive. I think it *is* subversive, but only because of the context."

A Grand Exbeeriment

Jeff Mello was ready for a change of career. His job at a Washington, DC, consultancy had turned into doldrums, so in 2013 he made the sensible choice to quit, and start brewing beer. The way he went about it, though, was a bit unorthodox. Picking up a trio of glass jars filled with unfermented wort—the sugary broth that yeast converts into bubbly beer—Mello walked into the backyard of his Arlington, Virginia, home, set them down, and walked back inside.

Mello's seemingly absurd hope was that at least one of the jars would, by dint of pure luck, produce a halfway decent beer. Absurd or not, that's just what happened. The next morning, Mello gathered the jars and set them in his closet to brew. After a few weeks, Mello said, "They started growing all kinds of interesting things in them." Mustering the moxie to stick his nose past the lip of the jars, he discovered that "one of them kind of had a really nice, funny, floral quality, and I thought there was some potential there."

Indeed, there was more potential than he could have known at the time. The source of that funny, floral smell was yeast that had wafted through the cheesecloth, which would ultimately carry him into the 3,600-square-foot facilities of Bootleg Biology, where we stood chatting.

Within a year of capturing the free-floating strain of beerworthy yeast, Bootleg Biology was open for business. There wasn't much to open back then, besides the door to Mello's kitchen, or perhaps the petri dish in which he kept that first fortuitous fungus, since dubbed *S. arlingtonesis*. With the most rudimentary lab skills—using a paper clip to scrape the sample across homemade agar—Mello had isolated *Saccharomycetes*, the most common genus of beer-making yeasts. Feeding it ever more wort, he worked it up to a pitch of viable volume for brewing beer. From then on, Mello was a committed yeast wrangler.

"I was kind of hooked from that point," he told me. "If I can pull yeast from the air, and use that to make a beer, and it's one hundred percent something that I've created, it just kind of had a magical quality about it, even if I was using some scientific methods to source it."

Beer is, at a basic level, fermented wort. The sugary brew of hops and barley make for a concoction that yeasts can't resist. Their voracious consumptive and replicative activity converts sugars into alcohol and, depending on the strain or species, imparts flavors that are described in wonderfully evocative terms of art: biscuitty, hayloft, horse blanket, and my personal favorite: goaty. ("I don't know many people who have interacted with a horse blanket," noted Mello.)

S. cerevisiae is the workhorse species of brewers (and bakers) the world over, responsible for the "clean" flavor of porters and IPAs (the sharp flavor of which comes from the hops); *S. pastorianus* are the lower-temperature "bugs" (the industry term for yeast) used for lagers. On the other end of the flavor spectrum are the *Brettanomyces*, or "bretts," widely considered a contaminant in brewing for the "horse blanket" notes they produce. However, the "off" flavors generated by bretts can be exciting for those with adventurous palettes, showing up in Belgian lambics for reasons we'll get into shortly. Certain bacteria, too, like *Pediococcus* and *Lactobacillus*, can be employed to add new notes to a brew. At that point, you've crossed into the frontier of flavors that craft brewers have made a mission of exploring.

"We call it funk, but it's earthy, leathery, barnyardy flavor," said Bryan Heit, a home brewer for twenty-five years who's also taught DIY yeast-culturing techniques as Sui Generis Brewing. "They also tend to have a lot of fruit character, often sort of apple, or plum, or raisin. It depends on the strain. The major difference often between a good strain and a bad strain is that the good strains will have a restrained amount of funk, so it's actually pleasant, whereas the bad ones, I mean, I've had some that basically tasted like burnt Band-Aids and soil."

Over time, different species and strains of yeast have been bred and domesticated, turned into reliable cultures that produce consistent batches. Massive yeast labs, like White Labs or Wyeast, produce pure strains at industrial scales, selling them as pitch to brewers of all sizes, all over the world. At these scales, the goal is less about flavor experimentation than reliability and consistency. For a brewing company trying to keep a customer base, that's key.

"There's a lot of really usable wild yeasts out there," said Mello. "The problem is that domesticated yeasts are so much more efficient. They're kind of like little machines that have been created by humans to efficiently make beer. So if you're a brewery that's on a timetable, or trying to create the same beer over and over again, it may not be ideal. But more and more breweries are trying to show that they care about making food that's as local as possible."

What Mello did in his backyard is commonly called wild fermentation, brewing beer from free-floating "bugs." Beer has been made for centuries by casting for ambient yeasts. Traditional Belgian lambics are made by way of a process called spontaneous fermentation, a term that suggests the time before living yeasts were identified as the cause of fermentation. Before then, it just seemed to *happen*. Same for farmhouse and saison varieties. In spontaneous fermentation, wort is left out in a massive "coolship," which basically amounts to a wide, flat steel vat.[26] Over the course of a few days, the local yeast and bacterial communities drift into the thinly spread wort, and inoculate it with the stamp of local microbial terroir. Of course, in place where beer has been brewed this way for generations, even centuries, the local microbial populations are likely to be favorable for beer making.

The first-ever ferments were all just the result of microbes drifting on the breeze, landing and proliferating in a puddle of honey or on a piece of fruit. In South America, chicha is made by masticating grains, fruit, or maize, which turns into a sour sort of beer when fermented with the microbes in someone's mouth. The slight dusting on the skin of an apple still attached to its tree or a grape on the vine is a waiting community of yeast, the agents that will, if properly managed, turn the fruit into cider or wine.

Sadly, beer doesn't grow on trees. The raw ingredients that ultimately ferment to make beer only come together under the guidance of human beings, and so the yeast that will carry the process forward to its final form must either be introduced under controlled conditions, or waft in with the wind. The odds of the latter making a good beer are generally lower. "When we're talking about wild yeast, we're looking for something that makes good beer," said Sam Wineka, Bootleg's lab manager and designated bioprospector. "Most yeast doesn't."

————

In the lobby of Bootleg Biology's Nashville headquarters, a table was draped with the company's logo: a spurred cowboy boot gingerly kicking an Erlenmeyer flask. It hasn't changed since the beginning, even if Mello's attitudes about science have.

"I think this is in some ways like a snapshot of my mentality when we first started," he said, with a hand placed thoughtfully to his long, brewer-worthy beard. "It's not a middle finger to convention, but it's saying that we're going to do things a little bit differently. And my perception of science at that time was that it was very rigid and maybe a little stuffy, and to use maybe a craft beer mindset—I don't know how you'd want to define that, maybe a little bit of irreverence—and bring that into science, at least as it relates to yeast."

Mello and his wife, Erin, had relocated to Tennessee to escape the Capitol region lifestyle; for Jeff, it was also a chance to be closer to a hub of craft beer culture. At the same time, Bootleg graduated from the kitchen to the basement, and ultimately to the facility in Nashville, where a staff of six employees now helps manage a library of yeast strains from near and far, for beer and for the sake of experimentation.

What started as an effort to source and save microbial diversity, and to promote the accessibility of basic science, has evolved into a bona fide yeast business. But that hasn't changed the original mission too much.

On the wall was a map of the world with every bit of land covered up, save for a handful of countries and a smattering of states. The uncovered spots indicated where a yeast strain in their library had come from, representing another part of Bootleg's mission: the Local Yeast Project. The project's half-serious original goal was to collect a strain of yeast from every ZIP code in the country, a range that has since expanded to the entire planet. It might as well be the whole solar system, considering there are more than forty-two thousand ZIP codes in the United States; at the time of my visit, they had banked just over two hundred.

Part of Mello's original inspiration to explore the possibilities of native microbes came from, appropriately, Sandor Katz, by way of his book *Wild Fermentation*. The invitation to revel in microbial richness rather than working to hinder it, struck Mello as deeply compatible with the craft brewer mentality. "They're more willing to take risks, I'd say, whether it's throwing Cocoa Pebbles in a beer or taking a chance on a yeast culture that doesn't have a proven track record," he said. "I feel like Sandor is willing to take a chance on wild fermentation, which has no guarantees except for your own experience. And I think taking out the question marks in wild fermentation, other than being a philosophical inspiration, shows that you can remove some of the more intimidating aspects of wild fermentation and have something close to consistency."

The impetus to actually pick up a jar and sterilize a paper clip and hunt brewable yeast, though, came by way of an article by homebrew guru Michael Tonsmeire, known as the Mad Fermentationist.[27] In it, he outlined the basic techniques for capturing, isolating, and brewing with yeast sourced from pretty much anywhere: the air, flowers, a doorknob. The methods involved were relatively simple. First, gather a sample, maybe with a Q-tip, or in a "trap" like the one Mello had set up in his backyard, luring microbes with the promise of a sugary meal. From there, it was just a matter of getting the captured community of yeast and other microbes to grow on a medium like agar, then isolate the desired organisms and keep them growing. Using the right agar limits the kinds of microbes that can grow, and narrows the search. The techniques are easy enough to grasp, but finding a good strain of yeast takes luck.

Bootleg also distributes "yeast wrangling" kits, containing the basic materials to carry out this process—petri dishes and agar mix, plastic pipettes, glass jar, a paper clip—the idea being that anyone who orders one can sample their own local microbial landscape and, if they find something worth keeping, send it to Bootleg for storage and, possibly, future use as a beer. None of LYP strains had yet made the cut for beer pitch, though.

Bootleg still sells the *S. arlingtonesis* strain from Mello's backyard, but the only other wild-caught strain among the thirteen they sell to brewers was called *S. cerevisiae boulardii*, which came from chardonnay grapes plucked at a winery in Washington State. The rest, like its Funk

Weapon series, are either variations on standard strains, or mixed cultures, combined and fermented in a test brewery at the back of the shop. The process is largely one of trial and error; try a mix of microbes, brew it, keep or discard to taste.

"Wild fermentation is great, but the controls that you have are the ones that are in your growth media," said Mello. "With a lab like ours, we can work with multiple pure yeast cultures, and use the 'canvas' of that beer, but then 'paint' in microbes, yeast and bacteria, and come up with a final 'work of beer' that didn't exist before, because we've in some ways unnaturally brought together microbes that have never been together before."

In brewing as in the arts, tastes are constantly shifting. At the time we talked, Mello noted that the trend in beer had been swinging away from "funky" and toward "clean," the quality most associated with lagers. A variety of fast-acting, very clean strains of Norwegian yeasts called kveik were gaining popularity. Genetically distinct from the traditional brewer's varieties,[28] kveik yeast was traditionally stored on a "kveik ring," a wooden assemblage that resembled a spine folded into a circle buried for safekeeping in stable subterranean temperature, or soaked into a cloth and then dried, like Sandor Katz's yogurt strain. Bootleg's own Oslo strain had been its first big international hit, to the extent that people were copying and selling it as their own, even in Norway. The strain represented Bootleg's intellectual property, but there wasn't much Mello could do about that. He didn't seem to mind all that much anyway.

"I think of it like we've accomplished something," said Mello. "People in another country are trying to repurpose something that we've done and think it has value, which is, I would say, leagues beyond anything I ever conceived would ever happen when I was working from my kitchen."

Bootleg's lab space was about the size of a small bedroom. An M. C. Escher lizard print on the concrete floor was left behind by the previous renter of the building, and Mello had sensibly declined their offer to remove it. Of more practical value were the ultraviolet sterilizer, flow hood, and microscopes that allowed for the accurate assessment of cultures. Some companies post their first dollar to the wall; Mello had posted his first agar plate. The portable Bunsen burner he'd employed in the kitchen days was still in use, a lab favorite. "I'll bet our first flask is floating around someplace."

Many of the same techniques Mello had used in his kitchen were still in play, too, albeit with a degree more sophistication. After streaking a culture across an agar plate, it could be visually assessed for purity. If allowed to grow, yeast and other microbes exhibit distinctive features on agar that skilled cultivators can assess at a glance. From a minifridge-sized incubator, Mello pulled out a plate and removed the lid to reveal the microbial Morse code inside, little strings of beige blips. Kept at a steady 35°C, each dot represented communities of pure yeast. Instantly, the faint smell of beer hit my nose, although really I was smelling fungi.

Sensory skills retain their value even in high-tech laboratories; beyond aiding in identification, they lend an aesthetic dimension that many scientists find an essential component of the work. But when selling microbial cultures for propagation in food and drink, the standards are rightly quite higher than sniffing petri dishes or jars of wort in the closet. The strains have to be pure and healthy, and contain exactly what the label says. So there are more sophisticated tools involved as well, such as PCR sequencing, which enables the Bootleg crew to verify the purity of cultures, as well as to measure cell counts, especially important when making pitch that's fit for reliable brewing. They also offer these services to other brewers and yeast wranglers. In a certain twist of irony, Bootleg Biology has become a resource of scientific rigor for those brewers who lack the lab skills Mello developed with his DIY attitude. The science of brewing, it turns out, much like the art, comes down to experience.

"Firsthand experience is as valuable to me as a diploma on the wall," said Mello, closing the petri dish and placing it back into the incubator. "I think good science operates in a gray area sometimes; you can't remove the human element of experience, which is probably the best thing we offer our customers. They rely on knowledge and expertise because not everything is a positive or negative result."

From the lab, Mello took me to see the culture library at the heart of Bootleg, which was located in the back of the building, a large converted garage. It lived in what was essentially a large ice cream cooler pushed up against the wall, dramatically gushing vapor clouds of condensation as he cracked the door. When the miasma cleared, there was a small stack of cartridges and a whole lot of empty space. It turned out yeast strains didn't take up much room. Suspended in hypersleep inside the tube marked *22204* was the first culture from Mello's backyard in Arlington, named for the ZIP code from which it came.

The freezer means that Bootleg can keep their cultures indefinitely in little plastic wells—each can be dipped into dozens of times before it runs out, and if that is about to happen, they can just toss in a little more wort to give the yeast a kick of life. Prior to getting the freezer, keeping cultures entailed restreaking plates over and over again, because of the instability of agar over time. Down the wall, a large flow hood facilitates the inoculation of test batches without worrying about contamination.

"I would say getting a freezer and getting a hood made us a professional lab," said Mello. "After that we could do anything; we may not have been able to scale very large, but it was the minimum to do exactly what we wanted to do, and it's good forever."

The company's own cultures represented just over a dozen of the wells in the freezer. The rest were from the Local Yeast Project, or stored as a service for other brewers, who often don't have the means of (or interest in) managing their own culture libraries. In launching a yeast lab, Mello had visions of fostering a culture of DIY science; a stand against propriety in the form of enabling anybody to source a key ingredient for their brewing. It turned out that wasn't a sustainable business model.

"I started doing Bootleg with the idea of open source: showing the world what your processes are, how do you get this into as many people's hands as possible, but then the reality of running a business is, not everybody subscribes to that viewpoint," said Mello. "Once you start getting to the point where you're like, man, we've got to protect our ability to operate as a business, and make sure we always have revenue to pay people and pay our rent. That's when you start thinking, not as a cold hard capitalist, but you start thinking about what is shareable and what's not."

"There's a lot of books about creating craft breweries, there's no books about creating yeast labs," he added, surveying the business he'd built. "It feels good to me, for Bootleg, because I was just a home brewer; there was no reason why anyone should have believed anything I said. So hopefully you've done things right when people trust what you're doing."

CHAPTER NINE

Bio Prospects

On a hazy August afternoon, a troupe of excited twentysomethings fanned out into the muggy shade of oaks and hemlocks in the woods of central Pennsylvania. Speaking sparingly, eyes fixed to the ground, they traced the mossy, wooded flank of a tributary to the Susquehanna River.

As I followed along on my first-ever *Cordyceps* hunt, it was presumptuous to even hope to be the first to spot the rare fungus we sought. Yet less than fifteen minutes into our foray, a slender orange mushroom in a mossy ditch caught my eye, and I called out to the group leader. William Padilla-Brown didn't need to get close to see that, unsurprisingly, I had struck upon fungal fool's gold: a cinnabar chanterelle. "The colors look very similar, especially this time of year," he said graciously. After a moment's more attention, even I could tell it looked nothing like the *C. militaris* we sought; for self-esteem's sake, I squinted at the mushroom in hopes of seeing more of a resemblance, but to no avail.

"You really have to get in here and tune in," he said. "It's a very refined situation that you have to create for yourself." The twenty-five-year-old mycologist then disappeared between the trees to regroup with his companions. Not two minutes had passed before a triumphant shout reverberated from down the creek. "Yooo! *Cordyceeeps*! Woohoo!" Scrambling over logs and ditches, I discovered the group in a crouched huddle along an embankment beside the creek. Edging in, it took nearly a minute before I could see the fungus at the center of all the fuss, a tiny ocher apostrophe hovering above the wet underbrush. Our companion Jacob had found the little thing, no larger than a baby's pinkie, that marked the troop's first big find of the day. The collective energy was suddenly charged and borderline ecstatic.

Cordyceps (cordies or cheetos, to those in the know) are subjects of intense passion for a growing number of young mycophiles across North America. They're at the center of a growing community of independent, tinkering cultivators, and a boutique nutraceutical economy in which Padilla-Brown and his cohort represent the DIY cutting edge.

Another reason for the rising profile of these fungi is the shocking lifestyle they've evolved. *Cordyceps* are entomopathogenic, meaning they kill insects. But they don't *just* kill them. One of the scenarios most often recounted plays out like a scene from *The Body Snatchers*. It starts when the spores of a certain species of *Cordyceps* take root in the carapace of an ant—different species target different insects. Hyphae then thread throughout the insect's tiny body, eventually seizing control of its nervous system. The ant becomes, in effect, a living zombie, unwittingly stumbling up a nearby branch, inevitably one that sits directly over the path most used by its hive mates.[1] There, its final, irresistible impulse is to latch its jaws upon the twig, dying as the mycelium finally consumes all the insect's innards. After that comes the unsettling coda; out of the back of the ant's tiny neck slithers a slender "stroma," its surface bristling and primed to rain spores down upon the next group of unfortunate ants below.[2] The species of *Cordyceps* found under this particular fern, though, targeted not ants but moth pupae.

"Typically the bug is not going to be less than a couple centimeters below, so I can tell pretty much exactly where this insect is right now," Padilla-Brown said coolly, peering into the small cavity beneath the foliage. Everyone else had gone silent, and for a moment the mossy microclimate of that small square foot of forest floor had become a zone of reverence.

In harvesting the fungus, there was a risk of damaging or breaking off the insect from which it grew; pickers take pains to avoid severing mushroom from bug so as to collect as much of the mycelium as possible.

A bystander offered his knife, but Padilla-Brown opted to use his fingers. The sandy soil offered little resistance. "Sometimes you'll be within the root zone of a tree, and you don't want to nick it with a knife, because you might get some kind of fungal contamination into it," he said. "Sustainable, ethical harvest of *Cordyceps militaris*. Again, a very refined thing."

The unfortunate arthropod finally emerged, and Padilla-Brown raised it up slowly for display. Between his tattooed fingers was a wad of dirt that roughly traced the shape of the pupa inside, a one-inch, two-tone stroma of *C. militaris* poking straight out from the top. There were big plans in store for this little fungus. As people maneuvered for a better look, someone asked what it was good for. "This has incredible energetic properties," Padilla-Brown answered. "It's really good for your immune system. It's really good for prostate health and sexual function. There are many desirable traits to this mushroom."

By the end of our two-hour hike, the specimen was joined in its plastic storage box by nearly a dozen others, separated by dividers to keep their spores from intermingling. After a tick check and a group debrief, we caravanned to the campground where the MycoSymbiotics Festival was carrying on.

Back at the campgrounds as dusk fell, the festival was shifting into revelry. A conspicuously young, diverse crowd was milling about the lawn outside the central hut, where farmer and educator Matt Powers was concluding a lecture on the intersections of mycology and permaculture—to rapturous applause. A pair of big green picnic tables were almost invisible underneath the clutter of mushrooms gathered from the dense, humid woods all around us. Even the groundskeeper swung by carrying a tin bucket overflowing with neon orange chicken of the woods. Here and there were substrate blocks, from which arose spectacular tangles of lacquered, candy corn–colored antlers, the work of a cultivator who had made an artistic medium out of designer strains of *Ganoderma* fungi. As night descended, a progressive rock band took the stage, and their audience danced euphorically in a raging rainstorm.

MycoFest is Padilla-Brown's brainchild. In its fifth year, it was still a smaller festival than many, but much larger than when it began. Some two hundred people crowded the campgrounds during the peak of attendance on a Saturday; in 2018, the number had been seventy-five. "It's crazy, man," he said. "I started this festival with, like, a hundred dollars. The first venue I had was free."

Festivalgoers hailed from all over the country, bringing a range of perspectives on the social, economic, and ecological opportunities posed by fungi. Phillip Balke of the Central Texas Mycological Society spoke about farmer co-ops and sustainable local economies, laced with Marxist undertones (slyly citing *Cordyceps* as an example of "seizing the means"); Tess Burzynski of Fungi Freights in Detroit spoke on the role of mushrooms in fostering urban food security, and of her work converting shipping containers into mycological resource centers; from South Carolina, mycological innovator Tradd Cotter delivered an impassioned speech about the future of medicinal and agricultural fungi, its tone soon resembling that of a sermon and inspiring raucous hoots from the audience.

By eyeball estimate, roughly half those in attendance were people of color. Most were quite young, too, to the extent that, being in my midthirties, I felt conspicuously old. It turned out many had come from surrounding urban areas, including New York City. "There are some people that come out from the city that literally are, like, 'I've never camped, I've never been somewhere where it's dark at nighttime,'" said Padilla-Brown. "I try and make sure I stick close to the city, so people can get all sorts of public transportation to get over here, and then just make sure there's things that attract a diverse crowd, like the music and art and all of that."

MycoFest featured many of the activities and subjects familiar to fungi-focused events: organized forays, lectures and panel discussions about mycoremediation, medicinal mushrooms, permaculture, ethnomycology. But at its heart was an intersection of culture and commerce, revolving around not just fungi but also growing food in general, and applying permaculture principles beyond the traditional contexts of farming and land stewardship. It was about forming sustainable, regenerative communities, based on encouraging region-centered activities and economies that put into practice at a social level the many "loop closing" principles articulated by permaculture and indeed embodied by fungi.

It was all part of a larger project that Padilla-Brown has described as social entrepreneurship, one of several names that emerge in trying to describe the scope of his ambitious, multifaceted project as a citizen scientist, farmer, educator, and culture creator. "Social permaculture" was another, or the mouthful, "ecologically regenerative sustainable microindustries," a term Padilla-Brown seized upon after attending a course by the same name. "Everybody thinks permaculture is an agricultural thing," he said. "It's way more than that. I was trying to get to a point of designing culture, and designing relationships, and designing community."

If Padilla-Brown has an official catchphrase, it might be "propagate and myceliate," a mycological metaphor for purposefully cultivating social connections and spreading ideas through communities of practice and cultures of exchange. At MycoFest, the information being transmitted —about growing, understanding, and identifying mushrooms, using their value as medicine and food, as the basis of local economies—was set in the context of a boutique product space. Tinctures, whole mushrooms, and various value-added products, along with books, courses, and other educational material, were part of a collective effort to kick-start local economies that provide goods and services, while keeping the exchange bound as much as possible to a community and its geographical area.

Small-scale mushroom cultivation offers a fine example of how such a system could look. Mushrooms grown by a local cultivator will likely be sold to their neighbors, more or less nearby, depending on the density of the area. In part that's because of the hard realities of scaling up, and also the shelf stability of the mushrooms themselves. As we've seen, many mushrooms do not travel well, and may be grown from strains sourced directly from the surrounding landscape, on substrates most likely sourced from local agricultural waste streams; one isn't likely to be importing straw from the other side of the country. The waste product of the growing process, too, can be applied to building local soils, and perhaps feeding into other virtuous cycles involving nearby farmers or local landscapers. For a time, Padilla-Brown himself grew and sold culinary and medicinal mushrooms, as much as two hundred pounds a week. At its peak, he said, his home-based business was the only mushroom farm within a hundred miles that was not one of the giant Kennett Square companies.

Meanwhile, cultivators focus on educating their communities about mushrooms not just to turn people on to how neat fungi are, but because doing so creates a wider consumer base and opens up opportunities for exchange and innovation. By sharing knowledge and skills, participants in a local peer-to-peer economy can serve the needs of their community rather than concerning themselves with scale or guarding their corner of a market. Bounded economies like these tend to be geographically isolated from one another to a greater or lesser degree, which keeps value—in this case, money, food, medicine, and skills—circulating within a region. It's a throwback to the model of the bazaar, "where people could exchange not only their goods but also their ideas, leading to innovations in milling, fabrication, and finance," as media theorist Douglas Rushkoff describes it.[3]

At MycoFest, a circle of vendors was set up around the central hall, and visitors perused the various booths, chatting with the proprietors and apparently spending a fair amount of money. The lion's mane–infused cold brew sent by the up-and-coming European mushroom supplement company Four Sigmatic was sold out early on. Olga Tzogas was also on hand with her Smugtown booth, books and pamphlets about anti-fracking activism sharing space with her jars of grain spawn and handkerchiefs with mushrooms on them. She had helped to organize the first three MycoFests; a fixture in the community with a steady stream of visitors to her table. In speaking with Tzogas, I was taken back to sitting at her Rochester warehouse, trying to understand how running a business made sense with community, rather than profit, as the organizing principle.

"These people make it go around," Padilla-Brown told me. "These people are the people that buy our products. These people are the people that come to our classes. I need that, I need that community because they put food on my plate and they put food in my kid's mouth."

Padilla-Brown is a bona fide mycological influencer, his profile steadily growing beyond the realm of fungi. He has tens of thousands of followers on social media, profiled in a growing list of articles and documentaries. He's an author as well, with two well-received books on the cultivation of *C. militaris*, including the first on the subject written in

English, and a seemingly endless list of ancillary projects that orbit the central goal of demonstrating and encouraging circular, local, sustainable models for small-scale agriculture. On top of that, he is also a father, and maintains a parallel career as a hip-hop artist, under the moniker It's Cosmic. One of his songs includes the lyric: "I just had an epiphany / Monetize myself be my own industry / I'm in this naturally, synergy / They just wanna snatch some of my energy." In 2016, he lived in the town of New Cumberland, Pennsylvania, where he ran for mayor in a campaign that he suspended for personal reasons, though he's quite confident he would have won.

Despite these accomplishments, Padilla-Brown has no formal education in mycology, ecology, economics, politics, nor any other field one might expect given the scope of his work. He describes himself as a "graduate of Google Scholar." The son of parents in the US Army and the Department of Agriculture Foreign Agricultural Service, at various points in his childhood Padilla-Brown found himself living in London, Mexico City, and Taipei, before ultimately settling in Pennsylvania. Moving from state to state, country to country, school to school, resulted in a fragmented education. Ultimately, after he dropped out of high school at age sixteen, he spent the next two years "focused inwards," meditating, traveling, spending time in nature, and experimenting with mind-altering substances.

"I'd be going outside and walking a lot, and getting visuals and things, but I realized a lot of the visuals were just the understanding of natural patterns," he said. "Once I understood it, I would go to some other level. And I was, like, wow, I can really understand the way nature works, maybe I should be working with nature."

Having grown up largely in urban environments, seeing trees as just "green blurs alongside the highway," Padilla-Brown first undertook the study of permaculture. A preternatural autodidacticism helped him advance his skills and knowledge in spite of a disjointed educational experience. For a time, he became enamored by the cannabis industry, the niche economies and culture of experimentation that formed around quality products, and the inherent potential for sustainable land stewardship, all of which has informed what he hopes to accomplish with fungi.

At age eighteen, Padilla-Brown began growing mushrooms. Two years after that, and shortly after the birth of his son, Leo, he committed to making a living in pursuit of mycology and his nascent concepts of social permaculture. "I was, like, I can't keep working as a server. I couldn't get good jobs, because I was a high school dropout. I had [Leo] when I was twenty, and I was, like, I'm not going to let myself fall into this stereotypical 'young Black male high school dropout with a kid that can't get a good job.' So I was, like, screw it. I quit my job the year he was born, and I was, like, I'm going all in." The gambit paid off, as Padilla-Brown finds himself in demand as a speaker and the increasingly familiar face of young, diverse eco-entrepreneurship. Though he's wary of tokenism in some spaces where he's invited to teach or speak, his message since has remained one of, in essence, "If I can do this, anyone can, and more people should."

Padilla-Brown's parallel preoccupation with fungi has for some time been spirulina. The blue-green cyanobacteria is nutritionally potent and prolific, replicating incredibly fast with little more input than water, carbon dioxide, and sunlight. It's among the oldest life-forms still on earth, and the subject of great excitement in conversations about the future of food.[4] Padilla-Brown is one of a growing number of advocates who—in addition to being fascinated by spirulina's nutritional value, incredible rate of reproduction, and century-spanning cultural history—see immense economic and agricultural promise in its integration with today's local food systems. That includes mushroom farming. Spirulina thrive on carbon dioxide, a fungal by-product, and create oxygen, which mushrooms require; imagine a system where the one exhales what the other inhales and vice versa, each sustaining the other in a literal mycosymbiosis.

For years, Padilla-Brown worked to establish circular, regenerative systems at his Pennsylvania home, much to the chagrin of the landlord. Gradually the scale and variety of interconnected projects became untenable. "I had this vision of complete self-sustainability," he told me. "I was able to grow and forage everything that we ate; I was fermenting things; I was drying things; I was saving my seeds, all of it. I got to the point where I was producing fiber to make my own clothes. I was making biofuels in my backyard. I could feed the spirulina to sea monkeys; I could feed the sea monkeys to small fish. And then I got these things called *Triops*—they're similar to sea monkeys in that their eggs can be dried up and reanimated—and I would use those to eat the mosquito larvae out of my rainwater barrels. It was just levels on levels of stuff, I was getting really far out."

From that point, he began focusing less on trying to build fully sustainable systems into his own life, and instead to encourage their formation around him. "I realized at some point there is no such thing as self-sustainability; there always has to be a community. But I wanted to take my community to be an example for what the world can be, what the future could look like."

———————

There are more than four hundred species of *Cordyceps*, each associated with a specific insect: spiders, grasshoppers, wasps, to name a few.[5] In Tibet, *Ophiocordyceps sinensis* and its host moth larvae are methodically plucked from the foothills of the Himalayas. Locally known as yarza günbu, or "winter worm, summer grass," it is regarded as a potent aphrodisiac, often fetching a higher price than gold, always with the insect still attached.[6]

In North America, alongside mushrooms like lion's mane, maitake, turkey tail, and Chaga, *Cordyceps* has emerged at the center of a fast-growing domestic market for medicinal fungi, representing an industry that is expected to exceed $50 billion by 2025.[7]

The medicinal benefits of *Cordyceps* are largely credited to a special compound it produces, called cordycepin.[8] The compound has been associated with anti-cancer, anti-fatigue, anti-inflammatory, immune and sexual function-boosting properties, among other benefits, elevating it to the realm of fungal superfood, complete with the full range of branding and value-added products that term implies. It's now easy to find *Cordyceps* coffee, butter, powder for adding to your smoothies, and as an ingredient in tinctures and extracts.

C. militaris was of particular interest to the MycoFest community, in particular its medicinal potential.[9] Ongoing research into medicinal mushrooms is promising, but as is often true of traditional and trendy medicines, not yet conclusive as far as the science is concerned.[10] Just as much as any physiological benefits, though, for the MycoFest set at least, they pose an opportunity to exercise medicinal sovereignty, and independence from pharmaceuticals, while serving as the basis of community-scale economies.

For the cultivator, it also poses an enticing challenge. *Cordyceps* aren't anywhere near so easy to grow as, say, oyster or shiitake mushrooms. Cultivation is carried out by large-scale facilities in China or other parts of Asia, which understandably do not go to great effort to share their methods with American growers. Materials documenting how to grow the recalcitrant orange fungus were essentially nonexistent in the English language until Padilla-Brown published his first cultivation guide in 2017. Since then, a fast-growing community of small-scale cultivators and genetic tinkerers has sprung up around the country and the world. "That was wild, that really changed a lot of stuff," Padilla-Brown said of his first book and the activity it has helped to inspire. Within a month of its publication, he said, people in more than twenty countries had bought the book; when we spoke, he'd sold almost five thousand copies. Padilla-Brown also appears to be the first in the country ever to grow the mushrooms commercially; meanwhile, others have launched their own businesses around cultivating the fungus, such as Mushroom Revival, currently the largest producer of *C. militaris* in the country. Before these developments, domestically grown, whole *Cordyceps* fruiting bodies were nearly impossible to find on the market, driving their import price as high as $120 and $100 per dry ounce.[11] Part of the goal in developing accessible cultivation methods was to get these prices down so that a domestic market could emerge—Padilla-Brown tends to sell whole mushrooms for five dollars a gram—which is indeed what appears to be happening.

The methods outlined in Padilla-Brown's book use brown rice as the substrate, a much more accessible—and less disturbing—method than inoculating living insects, a method often used overseas. But *Cordyceps* are more temperamental than many other types of marketable fungi in various ways. Cloning them is not as simple as with other mushrooms, as they are particularly sensitive to timing, light, and can breed into new strains that never actually form fruiting bodies, which is the whole point of cultivating them (although *Cordyceps* mycelium has also been used to make tempeh and other products that are, at least in theory, infused with their medicinal benefits). Breeding them generally requires either sequencing their DNA to find compatible mating types, or precisely isolating pairs of their threadlike ascospores under a microscope in order to get a reproductive match.[12] Most strains last about nine months before they senesce, and stop producing.

Developing the methods for propagating these temperamental mushrooms required an immense amount of trial and error, and Padilla-Brown didn't crack the code by himself. Cultivator and *Ganoderma* artist Ryan Paul Gates, whose fungus-antler sculptures were on display at MycoFest, spent many hours translating documents and videos from Thailand and elsewhere. With the decoded methods in hand, Padilla-Brown managed to coax his first *Cordyceps* to "pin," from a specimen discovered at the inaugural MycoFest in 2015, by Charlie Aller of the Mycelium Underground, who handed it over to Padilla-Brown in a gesture of camaraderie, and in hopes of seeing him succeed in cultivating it.

"It was the difficulty of it I think that created this hype around it, because people had been trying to do it for a long time," said Aller, whose nickname in the cordy community is Charlieceps. "I was personally moved by its medicinal value; I had an experience with Lyme disease, and I feel like *Cordyceps* was one of the linchpin organisms that brought me on home to some sort of health and normalcy. Once that had happened, I felt like I sort of owed it something beyond the more casual sort of reciprocal relationship of grower and consumer, organism and cultivator."

Our hike in the Pennsylvania woods represented what was called a "pheno hunt." Each new specimen uncovered in the woods would be brought straight into the basement lab, where *Cordyceps* obsessives would toil to tease out and propagate the most desirable traits. Some look for strange and interesting morphologies, or high yields and fast grow times; ideally they would also produce the most cordycepin, but to identify specific compounds required sophisticated biochemical analysis that few, if any, could perform or afford.

The designer strains that resulted from these field trips and selections formed the basis of an emerging economy of exchange, carried out at meetups, festivals, and of course online. Every day, more "cheeto" photos appear on mushroom Instagram, as experienced and emerging cultivators show off their successes. Tangles of bulbous orange, sometimes even deep red (the "flaming hot" variety, as they're sometimes called) compete for attention and draw the admiration of those tuned in to these efforts at pushing the genetic envelope. Gates is among the most sophisticated breeders, working with genetics sourced from all over the world, combining geographically distant strains to generate novel hybrids that may never have occurred in nature. The resulting strains have names like (Double Red #3 × (92b #6 × Cordyzilla #1316) #5) #5 × Mound4#4, which in this scene has definite meaning. Even without delving into genetic variations or painstakingly breeding ascospores—the methods for which are highly refined and often kept close to the vest—an enterprising cultivator can simply buy the strains to grow themselves, for personal use or to sell by weight, perhaps to someone who will incorporate them into chocolate or coffee beans or some other value-added product. They could also continue tinkering with the genetics, to push new and unusual varieties into the emerging market. There is precedent for this model in the cannabis industry, in which strains are refined to extremes of THC content, size, flavor, and other factors. "I have a theory that cordycepin is the new THC," said Gates in a podcast interview.[13]

Part of the reason for all this work is to generate a higher-value, affordable domestic option for medicinal mushrooms that are currently produced at industrial scales overseas. "There is no competition with China right now," said Padilla-Brown. "China produces eighty to ninety percent of the world's mushrooms, and they can give you mushrooms for pennies on the dollar. If you want to be able to produce mushrooms at anywhere near the price you get from China, you either have to have an automated farm run by robots, basically, or a massive outdoor thing with some sort of cheap labor."

There is debate, too, among medicinal mushroom purveyors as to whether the mycelium or the mushroom themselves are the more biochemically potent. Often, the dried powdered mushrooms one buys contain mostly mycelial fragments, including the growing substrate (grains, most of the time). The debate is known within the community as the "spore wars" and the two most prominent players are (you guessed it) Paul Stamets, representing team mycelium, and his former mushroom cultivation guide colleague, now professional rival, Jeff Chilton, who is firmly on team fruiting body.

The small-scale cultivators tend to be mum on their opinions in these matters, partly for reasons of politics, often opting for the "full spectrum" approach that says, in essence, why not use both mushroom *and* mycelium? Either way, making whole mushrooms available fills a real gap in the market. The dry powders are often highly processed, the result of hot-water extractions, followed by dehydration, powderization, and then spray drying. By the time they reach a customer, they don't resemble a mushroom at all, with largely unknown consequences for their potency.

Part of the hope in promoting ethically, sustainably sourced, and unmitigated mushrooms is that they'll be worth the extra price to conscious consumers. Here again the cannabis industry sets an example. High-quality extracts such as oils, concentrates, hash, shatter, and other refined forms are familiar to informed cannabis consumers. A very similar product class is beginning to emerge around *Cordyceps* and other medicinal mushrooms. The question remains open whether there is as high a ceiling in the market for medicinal fungi as there is for pain relief and recreation represented by CBD and THC.

———

As I descended into the basement of Padilla-Brown's Mechanicsburg, Pennsylvania, home, a pink glow gave the impression that I was entering a mad scientist's lair. That was, in fact, not far from the reality.

Hidden in the corner behind a zip-up air lock was Padilla-Brown's lab. Positive air pressure ensured that contaminants escaped, rather than entered, and a stereo made sure that he didn't lose his mind after a full day of staring into readouts and petri dishes. Multicolor agar plates were stacked in the corner, encrusted white with dried bleach, a measure taken against soil mites that often hitchhike on *Cordyceps*. "Don't touch that," he warned as he inoculated tubs of substrate, noticing my curiosity, "It'll make your finger feel weird."

In a small refrigerator in the corner, stacks of bleach-free plates contained the strains he'd been cloning; many were the result of previous phenohunts. After finding the mushrooms, he'd cloned them in the agar, and with a scalpel and sterilizer had propagated the resulting mycelia in dozens of glass jars stacked on racks outside the air lock.

"When you deal with all these clones, probably like twenty to fifty percent if you're lucky are going to produce mushrooms," he said. "A lot of them won't do anything. So the ones that produce mushrooms, I then took spores from those, and then I did breeding, which took a long time, and I had to figure out the whole DNA thing. I taught myself molecular biology in like two and a half months. I'd seen people do it a couple times, but I just watched a bunch of YouTube videos and asked Alan Rockefeller and asked Craig [Trester] questions, and ended up with good results."

A flow hood was set up next to a microscope. Beside that was a small thermocycler, used to prepare samples for genetic sequencing. Padilla-Brown conducted his own analyses to determine mating types, using primers ordered by mail and set to target the relevant genetic segments. He wouldn't need to go through the entire sequencing process, though. After preparing samples from a newly cloned specimen, he could visually assess their mating types using relatively simple gel electrophoresis—the process that separates genetic samples by length into easily read parallel rows, or "bands"—usually an intermediate step taken to verify sample quality before sequencing. The whole process took about two hours.

On the counter, I recognized the stapler-sized field sequencer I'd seen at the Natural History Museum of Utah. Padilla-Brown was excited about the possibilities posed by sequencing genetics in the field. He was looking to DNA to assess breeding compatibility, suggesting a wider range of fungal applications for the emerging sequencing technologies. In teaching these citizen science techniques, Padilla-Brown hopes others would take and run with them as he has. "My hope really is that people start working with this technology and figure out other stuff," he said.

"As far as mushrooms go, most people just do genetic identification, but there's so many other things we can do. It goes way beyond taxonomy; that's just what people focus on."

I had to take a mental step back and appreciate the scene. Here was someone without any formal scientific education, casually innovating genetic sequencing and cultivation techniques in his basement that not only produced interesting biological results, but also served as the basis of several small but successful businesses.

Exiting the air lock, we walked up to the source of the pink light. A series of warehouse racks upheld corridors of glass jars, the twisting orange fingers of *Cordyceps* growing atop cakes of myceliated rice. Held up by bungee cord and a stepladder, banks of LED lights provided the light wavelengths determined to maximize growth: red, pink, and blue, the latter of which encouraged the mushrooms to pin. Seeing all this, I had to ask, why *Cordyceps*? "Why *me*?" he retorted. "I didn't choose this."

Padilla-Brown wore sandals that revealed his toenails painted in vivid purple. He stepped over green stains on the cement floor, signs of where a spirulina experiment had gone wrong. Pressure cookers were strewn about, unused for some time. For the time being, he was focused more on writing the next cultivation book than producing mushrooms at any kind of economic scale. The mushrooms he was growing would be put to use by Cassandra Posey, his partner in both senses of the word.

Upstairs, we sat down for dinner, where Posey joined us to explain their business concepts, and goals for the future. Posey, whose company Cognitive Function made use of the *Cordyceps* grown in the basement as the basis of its line of tinctures, ghee, coffee, honey, and other fungi-infused products, had in turn helped improve the branding of Padilla-Brown's company, MycoSymbiotics. She had also taken the lead in organizing the previous MycoFest, creditable for the marked increase in its attendance and scope. It dawned on us both that I had bought *Cordyceps* coffee from her at the Radical Mycology Convergence in Oregon. Posey was hustling from coast to coast to get people interested in the "forest to table" health products produced by her, Padilla-Brown, and the growing community of eco-entrepreneurs in which they were becoming leaders.

"Will thinks very here and now, which is so great—I wish I could be more present in the moment," she said as we set down our forks. "But my parents heavily instilled a lot of the big-picture, five-year-plan kind of thing into me. I grew up watching my dad run a company, and all that is like my playground."

"I'm not a business person at all," added Padilla-Brown, pulling out a Nintendo Switch as our energy wound down with the evening. "I'm not good at business, I don't like doing business, I like doing whatever I want to do whenever I want to do it, which is beneficial for the students that I teach."

The couple, I realized, were a living example of brand synergy, finding a productive intersection between Posey's business acumen and Padilla-Brown's multifarious projects and experiments. They told me of a story from the time just before they'd started dating. Padilla-Brown had managed to ferment cacao beans with *Cordyceps* mycelium, a difficult trick given its temperamental nature. On a visit to New York, he presented them to Posey as a gift. "That's when I fell in love with him," she said, half jokingly. "He was, like, 'I brought you a present,' and then he hands me these freaking beans, and I'm just, like, 'I'm not going to date this guy, we're *just going to be friends.*'"

It was late by the time dinner ended, so I slept on the living room couch. The next morning, Posey brought to the table an intriguing, potent new extraction experiment the two had been working on in the kitchen, one that was to be kept secret, but which had everyone marveling at the product possibilities it represented. She also detailed a new business concept that, while also off the record, sounded as though it could effectively organize their nascent but fast-growing network of small-scale producers into a framework that might gain traction at a national scale. The scope and ambition of their plans suddenly seemed incongruous with the scale of the kitchen in which they were hatched and, to some extent, executed. Big ideas, though, like culture, emerge and spread in ways they didn't used to. All it would take is a post sent from one of their phones, perhaps a photo of some new product or experiment, to get the wheels turning and people engaged. In the kind of world they hoped to realize, perhaps success looks different from what we've been conditioned to expect. "Will likes to say that even if you have two followers, you're an influencer," said Posey.

Welcome to Mycotopia

Traveling to Mushroom Mountain was bound to feel like a pilgrimage, and not just because of its name. Mycologist Tradd Cotter's base of operations is an oasis for those of the fungal persuasion.

Nestled among the billowing hills of Pickens County, South Carolina, you know you've reached it when you see a large letterbox sign on the side of the road reading, WELCOME TO MYCOTOPIA. Atop the sign, a glowing triangle is decorated by a hand-drawn illustration in pastoral tie-dye colors, featuring morel, chanterelle, and porcini mushrooms sprouting along the sides of crumbling pyramids, and a gargantuan *Psilocybe cubensis* looming over all of it.

But you won't find any *Psilocybe* mushrooms there, heavens no. Cotter has another company for that, called Blue Portal, which at the time of my visit was reassessing after its plans to set up a psychedelic therapy center in Jamaica—a country where the blue-bruising mushrooms are totally legal—had fallen through. Mushroom Mountain has plenty else to keep it busy anyway. It is a research laboratory, a spawn and fresh mushroom producer, and proof of concept as an "amusement park for mushrooms," as the late mycologist Gary Lincoff described it. Every month, groups take guided tours of the place, although visits are also invited year-round, for workshops, shopping, or a wander along the old-growth mushroom trail. It's essentially a fungal take on Willy Wonka's Chocolate Factory, sans the Oompa Loompas and wallpaper anyone would want to lick.

Olga Cotter greeted me in the amply stocked gift shop that doubled as a meeting area, where we chatted while her husband, Tradd, was still en route. Along the walls were a trio of foggy-windowed coolers like ones you'd see in a bodega, filled not with beer but with grow bags, spawn, and whole mushrooms. I mentioned to Olga that a friend I'd be staying with that night had requested I pick up two distinct types of shiitake spawn, which she brought out from the back room as readily as if I'd ordered cold cuts at the deli.

Mushroom Mountain was officially founded in 1996, but it didn't start as a seventeen-acre fungal wonderland. "We started off in a little closet in our condo down in Florida," said Olga while running my card. She and Tradd had bought the property just five years prior, from a former importer of wicker baskets. A short walk up the hill was a huge warehouse stuffed with so many baskets that they still hadn't worked out a practicable solution for getting rid of them. Early on, Olga nixed Tradd's idea to simply incinerate the whole lot.

They would soon need to make a decision, though, as their business was growing fast. In just two and a half years, the number of employees at Mushroom Mountain had increased from one to nine. Among their top customers, Olga told me, were the monks at Mepkin Abbey, a nearby Trappist monastery that had taken to growing and selling mushrooms.[14] Besides spawn, another dimension of the business revolved around producing mushroom tinctures and extracts. Olga alluded to plans to enter the pet care market, using extracts of mushrooms commonly taken by cancer patients. A few of their friends' pets were already getting drops of the extracts in their dinner, part of an unofficial, very nonclinical trial.

After a few minutes of chitchat, Tradd made his entrance. "Look what I found," he said with nonchalance, as he dropped a heaping trash bag full of hemp onto the counter with a righteous crunch. The nose of a Shiba Inu named Zen, belonging to Olga's sister, snooped around the corner to hazard a sniff. Also salvaged from the side of the road was a lamp-sized, yellow-bellied bolete, tragically a few days past its prime. As staff took interest in the pungent bags, a playful dispute broke out over who would get which. "You see the hostility here?" he joked as we set out. "I thought CBD was supposed to calm you down!"

The standard tour of Mushroom Mountain starts at the back of the gift shop, which is also the entrance to the lab. Or rather, a place where you can see *into* the lab, *Jurassic Park*–style, through a pair of big windows. There, we stood and watched as a technician worked among all the standard trappings of flow hoods, sterilizers, racks of glass jars. But this was a lab with many uses, from developing strains for commercial cultivation to experimenting with their medicinal and remediation capacity. Certified as Biosafety Level 2, stringent enough for mildly dangerous pathogens, with hardware like a recirculation hood to protect workers from bacteria like *E. coli* and streptococcus, Cotter and his staff tested various fungal strains against these bacteria in what he describes as "gladiator matches" on petri dishes. "The ones we tested that were the highest against strep were reishi and chicken of the woods," he noted. "Jack-o'-lantern were higher against staph."

The lab is modular, so that equipment can be easily reorganized to accommodate cultivating mushrooms or performing research. It was also the first thing Tradd and Olga built, at a cost of about half a million dollars, testament to its central role to their work. Cotter is on a constant lookout for new strains and species that can fight bacteria, dissolve waste, or eliminate pests. At the time of my visit, the lab kept more than 250 species in its library, a collection that was growing at a steady rate. "We just keep bringing stuff in," he said. "Or people ship us things." Cotter's reputation as a fungal tinkerer means that people are constantly mailing him mushrooms they've found growing out of strange things. If a fungus can break down, say, a bowling ball, maybe it can break down lots of things made from the same material.

Cotter has a well-earned reputation as a mycological innovator. Clever applications for fungi seem to fall out of his pockets: inoculating wooden cutting boards with antimicrobial fungi that remain alive in the cracks, killing foodborne germs; cloning a gigantic *Macrocybe titans* to use as growable stools around the firepit; snagging fungus-killed bugs from the wild so that the entomopathogenic fungal species can be cultured and applied as nonchemical pesticides. He's also considered an authority on mycoremediation, the field of research that uses fungi to intercept toxins in contaminated soils or to filter contaminated waterways.

Next to the lab window, on a small shelf were some tangible examples of his fungal investigations. In a petri dish, a confrontation between the mycelia of two competing fungi, dehydrated and frozen in time like little waves crashing against each other. In a plastic sandwich bag, the first *Cordyceps*-killed ant that he ever found, kicking off his research into myco-pesticides. The tiny antlers of the *Cordyceps* stroma had long since broken off, the result of touching by too many curious hands. I had to comment on how much territory we'd already covered within the first stop. "That's why the tours are fun," he said as we walked on through the next doorway. "They just get to see all the uses. Hey, antibiotics! They can kill fire ants! They can make food! We can make soil; we can make building materials. Talk about psilocybin and it's, like, hey, there's stuff for your brain."

A wiry guy in his early forties, Cotter sported a crew cut dyed popsicle blue—"It just bruises that way!"—with a sartorial style that might be described as "tacticasual." Prone to quick-witted wisecracks deployed with a hint of Carolinian drawl, he is a gifted communicator, traveling the country to deliver energetic lectures that double as barn-burning performances, an echo of his previous life as the singer of a rock band. He consciously takes on the role of comic relief, drawing as much from Bill Hicks as from Bill Nye the Science Guy.

All the talk of fungi and mushroom cultivation is leavened with reflexive irreverence. In his slide show at the Telluride Mushroom Festival, viewers saw images of spore prints on a mirror, like a scene from Studio 54; photoshopped pictures of Donald Trump crawling along the White House lawn with a *Cordyceps* stroma sprouting from his head; a satirical title card for his speech flouted any authorities in the audience with a massive headline of: LARGE SCALE CULTIVATION OF MAGIC MUSH-ROOMS FOR FUN AND PROFIT. "I'm a mushroom missionary," he said later in that speech, betraying a deep well of sincerity. "And my mission is for mushrooms to take over the planet. And you know why? Because they're going to help us fix this place!" The night's presentation concluded with him stalking the stage in a creepy rabbit mask, to toy with anyone in the audience who happened to be tripping.

Raised in a military family, Cotter grew up in far-flung parts of the world, including Egypt (hence the pyramids in his company's hand-drawn logo). Although his interest in fungi came later in life, in retrospect, it seemed almost inevitable; in a picture he sometimes shares, his grade school soccer team stands gathered behind a placard showing its name: Fungus. Cotter had forgotten all about it until his dad discovered the photo. "He goes, 'Tradd. That is weird,'" Cotter recalled, before noting that it might not be quite so providential as it seems. "I remember some of the other kids, they were just making fun, going, 'We should call it Fungus!' Like a foot fungus."

Cotter's interest in mushrooms began with wild foraging. But his original career aspiration was altogether unrelated. He was on track to become a fighter pilot, a path he abandoned shortly before diving into mushrooms. It may explain why he sometimes describes his work with fungi in war room terms. He characterizes birds as fighter-bombers that disperse the spores he wants to proliferate in a landscape. He pictures rapid deployment of mobile mushroom-cultivation operations, established like prefabricated bases in the middle of food-scarce environments. On the Mushroom Mountain property, a nearly literal (if one-sided) war was being waged with fire ants. Tall red mounds used to ring the building that housed the lab and growing operation. After identifying a strain of *Cordyceps* that preyed upon them, he marshaled their spores to infiltrate the nests. The trick, he said, was to infect the queen; his interns had the unenviable task of digging for them.

"There's a couple now, but they took a hit, and I'll be honest, we didn't even spread the shit in here, it's just spreading naturally," he said. "Last year there were just dead ant bodies everywhere, it was awesome. There was just piles of them, like graveyards, you could scoop them up with a spoon. The guards were dragging them away as fast as they could.[15] Then, about a week later, that mound was gone."

Moving on from the lab, we entered a mushroom-cultivation operation unlike any I'd seen. The first thing that stood out was the first thing we saw: a custom-designed autoclave with a tapered chamber like a cement mixer, that sterilizes substrate and then allows it to be inoculated in the very same chamber, saving a big, time-consuming step. The spawn went straight from the autoclave into the bag, ready to incubate. It could reach twenty pounds of pressure, well above the standard for sterilization, a bespoke beast of a machine that I was very definitely not allowed to photograph. "That was needed to grow mushrooms, but it also makes a shitload of spawn," said Tradd. "That thing is surgical steel on the inside, it's mirrored, so I could use that for anything food-related that I want."

The next room, naturally, was the incubation chamber; spawn didn't need mixing in thanks to the NASA-grade autoclave, so the bags went straight to the racks, where the mycelia could run through their substrate. We passed by an entire pallet of bags stacked by Mushroom Mountain staff, and Tradd struggled to identify the species. "I have no idea," he said after a minute of leaning over and inspecting the bags. "That's what's great, they have so many things going."

Next came the fruiting room, and again I was taken aback by the unusual arrangement. Usually, mushroom blocks are stacked on shelving racks, or hung from hooks like giant strawsages as the mushrooms pop out. This arrangement instead positioned the bags on their sides, stacked as walls inside long, narrow wooden frames reminiscent of the old Connect Four tabletop game. The interesting part, though, we couldn't actually see. At every point where the bags made contact with one another, holes had been perforated in the surface so that the mycelium of each bag could connect to that of its neighbor. The wall of bags therefore formed one big, connected organism. Besides representing the single largest mass of fungus I'd ever (knowingly) seen, it was also prone to fruit in a single flush, as a big wall of mushrooms, instead of on a bag-by-bag basis.

"Everybody was upset with me for a little while because I didn't ask, I just did it," Cotter laughed. "I said, just trust me, *trust* me, and two weeks later, shit was like, whoa. We're doing three hundred fifty pounds a week, and we're not even trying."

As we meandered through the fruiting room, we looked in on a soil-building experiment using spent *Stropharia* substrate to feed red wriggler worms, which left behind a mass of pure worm castings, a prime organic fertilizer. There was a clunk and a hiss as the misting system roared to life, filling the heavily modified barn with fog. Further modifications were in the offing, including plans to build a twenty-by-forty-foot greenhouse against the back door and fill it with plants, to circulate air, much in the way William Padilla-Brown envisioned with spirulina (indeed, Cotter is a mentor to Padilla-Brown). The spirulina could then be thrown back into the autoclave as a nutritive supplement for growing mushrooms, closing a loop that might otherwise be waste. Another benefit of the greenhouse was that the plants would warm the air in the summer months, saving energy bills. Instead of a greenhouse, though, for the time being, they merely had a lone orchid hanging from the side of a rack.

Fungi seemed to factor perfectly into Cotter's mind for circular systems, something that expressed itself even in our own movements—always moving to the next room, never back, to prevent contamination. For a tinkerer, there was always some step to shorten, efficiencies to refine, and no chance to experiment was ever passed up. This was a working mushroom farm, but it was also meant to make an impression.

"I'd like people to come here and be, like, 'Oh that's innovative, that's crazy,'" said Cotter. "If they're, like, 'What's up with the plant?' And I say, 'Well, mushrooms make CO_2,' they go 'ooh,' and put it together. You can think of this like a little Mars module, that's basically what we're going for: closed systems."

Nearing the end of our tour, Cotter took me to a side room where visitors aren't typically invited. Sliding open a concealed door, we stepped down into a dark, brick-walled wing of the barn. It was unoccupied except for some plastic containers, a floor squeegee, and, tucked into the corner, three short, plastic troughs lined up side by side, with faucets at the top and buckets at the bottom. It was a mycofiltration demonstration room, still very much under construction, set up in preparation for a test to remove *E. coli* bacteria from a runoff site in nearby Greenville.[16]

In the basic version of mycofiltration, water passes through or over a mycelial mass; fungi are chosen for their ability to exude enzymes that reduce the target chemicals or microbes, degrading them of binding them up within the mycelium. In Cotter's test, each channel was filled with silt, with recirculating pumps to keep water flowing. The test would involve taking the strain of *E. coli* from the site, adding it to the water, introducing the fungus, and measuring how long it took for the bacteria to filter out. The fungus they had chosen for the job was a tiger sawgill, cultivated from a specimen sent in by Cotter's own mother.

"I ask people to turn things in, and she found it," he said, noting that the mycelia had shown qualities that made it ideal for this particular application. "When I was working with it in the lab, I noticed it had a really high tensile strength. A lot of these [fungus] filters were being blown out, they wouldn't hold themselves together, but that one was like glue."

Representatives from the city were scheduled to arrive the following week, and Cotter was confident the demonstration would be a success. After one test using a less resilient king *Stropharia* fungus, he said, they got a 100 to 1 reduction of *E. coli* in one hour. "These should do it within minutes, because it's tight-knitted," Cotter said. "We're going to try to blow them away."

Out the back, past a packing room, was a shipping dock where the spent substrate had been piled in massive heaps, sprouting huge flushes of mushrooms in the muddy yard. Operations had been slowly shifting back to growing mode, and Cotter mused with an employee over what to do about all the clearly still viable mushrooms that would soon go to waste. A consensus was reached that they would open a channel to a nearby soup kitchen; hopefully, another loop closed.

Walking from the substrate piles over to the trees, we entered the mushroom trail. Having spent an hour inside, it was a welcome shift of gears, and this is just where the tour group would wind up after their run through the farm. The trail meandered through the old-growth forests that abutted the property, with a creek running alongside it and some thirty-two different fungal genera planted throughout. Clusters of shiitakes were exploding from the stack of logs, but most of the other mushrooms were keeping to themselves during my visit. A little placard with a hand-painted portrait of the mushroom gave us something to look at anyway; for some reason, *Jurassic Park* again came to mind, but thankfully there was no need to worry about any stinkhorns tearing through an electric fence. The insects, though, might do well to be on guard. On a branch at the trail entrance, Cotter had tied a little flag to mark a special spot; another *Cordyceps*, so small it was almost invisible, sprouting from the head of an ant, its jaws clamped to a twig.

Outside and up a hill behind the converted barn complex, Cotter and I took a seat by a pizza oven that was—yes, of course—held together with heat-resistant mycelia. I felt compelled to ask him if, after twenty years of working as a "mushroom missionary," his situation represented a dream come true. "It's pretty wild," he said. "The weird part? I don't even feel, like, accomplished at all. If you would have shown me this, time traveled back and shown me this picture, I would have been, like, get the fuck out of here, no way. No way! And it just happened. So it's a gradual thing."

Five years prior, while Cotter was studying at nearby Clemson University, Mushroom Mountain had taken up a small room in his and Olga's home. Now they were reaching the capacity of their buildings, and contemplating what to do about the baskets as they nurtured plans for a Mushroom Mountain University.

In recent months Cotter had filed three trademarks on novel fungi-based disinfectants, and sent some thirty trademarks to his attorney. He'd quietly posted the newly trademarked products online for just one hour, before selling them to his sister for a dollar, and then took them down, to be able to show proof of use in commerce. It seemed that for all the dazzling displays of ingenuity, not unlike fungi themselves, the really interesting activity was taking place under the surface of what I'd seen. "Don't chase a dead rabbit," he said of his entrepreneurial approach. "Where are the other rabbits running?"

Mushrooms have been good to Cotter, sending him far and wide to speak and teach on their behalf, and affording opportunities to establish mushroom farms in Haiti, Jamaica, and other parts of the world in addition to his own business. For five years, he's taught the FDA Wild Mushroom Food Safety Certification Course, required in eight states— South Carolina, North Carolina, Georgia, Pennsylvania, Virginia, New Hampshire, West Virginia, and New York—in order to legally pick and sell wild mushrooms to restaurants. Originally uninterested in even *taking* the test, he was recruited to redesign and ultimately administer them. "We pulled it off, and they said, 'We want you to teach it,' and I said I don't have time. Then they showed me what it would pay, and I said, yeah, I've got time. I've got plenty of time."

Tradd credits Olga with running the business, seeing himself more as a product developer than a manager. And indeed, much of his work seems rooted in bringing people to imagine what beneficial applications might be possible with fungi. That vision includes food security, ecologically safe pest control, and local economies fueled by low-cost mushroom cultivation. One of the most enticing projects is the proposal to create "pharmacies in a bag": the mycelium of specially selected fungi growing in a sealed bag that, when injected with a sample from a human patient's infection, will in theory start producing enzymes tailored to fight a patient's specific infection.[17] It is exciting to imagine the possibilities of individually tuned antibiotics derived from living mycelium, but the testing of the idea's medical viability is ongoing. It's these kinds of unexpected ideas that the mushroom farm–amusement park hopes to spark. Perhaps some industrious visitor will go home with a notion that kicks off a kind of butterfly effect that helps change the world for the better. It's a hopeful thought.

The idea for Mushroom Mountain actually came, in part, by way of butterflies. Shortly after Tradd and Olga first met, they took a trip to a place called Butterfly World in Coconut Creek, Florida. The massive attraction takes up two giant geodesic domes in which the insects are unleashed in swarms that dazzle and delight visitors. Cotter was deeply moved by its example of nature education.

"The first thing you pass by is a lab," he said. "You see somebody actually working. It's not an actor, they're actually taking these little chrysalides from different species of caterpillars and butterflies, and they're all in jars. They have these little containers, and you see all these little pupae hanging in these little flasks, and you're, like, holy shit, what is that? And then as soon as you walk in the release room, there's *Heliconius* butterflies, and all these tropical butterflies, just clouding all over the place, and they'll land on you. And I was, like, this is awesome."

Cotter's voice took on an excited tone as he recounted the experience. "Now I've got goose bumps, see?" he said. "Because I want to do that with mushrooms."

CHAPTER TEN

APPLIED MYCOLOGY

O n June 8, 2002, at a campsite in the Pike National Forest about ninety-five miles southwest of Denver, a fire prevention technician for the US Forest Service did something unusual for her line of work, and started a forest fire. Flame is natural in forests, of course, and is frequently employed as a tool of forestry, but this was by no means a controlled burn. Over the course of six weeks, some 138,000 acres were blackened across four counties, leading to the loss of six hundred structures and causing more than $38 million in property losses alone.[1] Amid the chaos, an elderly woman died of smoke inhalation, and five firefighters dispatched from Oregon to help fight the inferno were killed en route in a car accident. The Hayman Fire would go down as the biggest blaze in Colorado's recorded history.[2]

The often-repeated narrative is that the ranger deliberately dropped matches into a faulty firepit, with plans to take credit for stopping the resulting blaze.[3] Whatever the motivation behind it, the immensity and intensity of the Hayman Fire were largely a function of the state of the forest itself.

In 2002, Colorado was in the midst of a drought, naturally increasing the odds of a conflagration. But another problem lay in the composition and distribution of the trees. To an untrained eye like mine, the mountains southwest of Denver present a faultlessly beautiful quilt of ponderosa, lodgepole, and other mixed conifers. To those who understand how fires behave, they represent a landscape of closely spaced matchsticks.

"When a fire occurs in a forest that looks like this, it will almost immediately become a crown fire," said Jeff Ravage, North Fork Watershed coordinator for the Coalition for the Upper South Platte, "which then will be driven by the winds twenty feet above the ground, move independently of the ground fire, travel thirty, forty miles an hour, and just rain down death as it goes. It's one hundred percent mortality, in many cases, for everything underneath it."

As long as there have been forests, fire has played an important role in keeping them healthy. But naturally occurring forest fires tend to be far less intense than the massive, headline-making infernos that can be seen from orbit. Instead of creating vast smoldering moonscapes, the more modest, natural fires of forests past merely cleared underbrush, churning carbon back into the soil and triggering a wave of new life on the forest floor. By opening new habitat space, such blazes encourage greater plant and animal diversity. Flame, like most things, is best in moderation, something Indigenous land stewards have known for millennia.[4]

I met Ravage at a shopping center in Conifer, just up in the foothills west of Denver. He arrived in the coalition's trusty pickup truck, dressed for the field in khakis and boots with his ponytail tucked underneath a baseball cap. Before venturing into the realm of forestry, Ravage had actually worked in web design and even had a career in Hollywood; his deep sense for the forest's flora, fauna, and microbial life made it clear he'd spent at least as much time among the trees as in front of a screen.

Ravage's job was essentially to manage CUSP's forest restoration efforts. That included mitigating damage by gradually returning tree populations to prehistoric conditions, in terms of species, spatial arrangement, and density, so that they are able to "carry" fires that don't turn catastrophic. This involved working with logging companies to remove and replant trees in a manner that preserved the seedlings and saplings awaiting their chance to grow. It also meant no more clear-cutting, and finding ways of encouraging biological activity in the soils around logging sites and burn areas alike. To do that, CUSP was turning to fungi.

"In the 1940s and '50s, if you were a forester, your job was basically working for a logging company," said Ravage, easing us up the mountain roads to our first destination. "You were trying to get the most board feet out of a hectare." At its peak in 1955, logging removed some ten million board feet from the Colorado State Forest.[5] Over decades, the densely packed, homogenous forests that resulted were managed so as to prioritize less frequent burns, rather than to be resilient to burning. Ironically, that meant a greater risk of ever more catastrophic fires, a legacy that lingers today in the form of thick, uniform stands. Add ever more dry conditions to the mix, and, sure enough, the frequency and intensity of fires have steadily increased.[6] The current picture of bygone forest composition came, appropriately enough, from old photographs—we were traversing the same landscape as the settlers that had arrived to colonize the area hundreds of years ago, but we saw a different forest.

"The forest used to be a mosaic, it used to be filled with meadows," Ravage explained. "The structure of the forest was created by the action of the fire. When a fire went through, and if it had an area of high mortality, the first thing that would return, assuming that you have the moisture, is aspen. And then the aspen groves would be infiltrated by conifers; the conifers would grow up and shade the aspen and kill it, and take over. And then fire will return." In this area, he said, fire should return between every fifteen to forty years, essentially turning over the understory and leaving the trees. In a catastrophic fire, by contrast, all that remains are bones. "There's no bacteria, there's no fungi in the soil, there's nothing but fused gravel because it's been through a pottery kiln." After a large-scale fire, it can take centuries for a forest ecosystem to bounce back.[7] With global warming exacerbating things, Ravage feared, they never would.

CUSP came together after the 1996 Buffalo Creek Fire, which burned 11,700 acres in the Upper South Platte, the largest fire in Colorado history at that time. The nonprofit makes its mission one of protecting the waterways of the Upper South Platte, a 2,600-square-mile range of mountains that channels some 80 percent of the water used by residents in and around Denver. Managing a forest watershed means keeping soils healthy and preventing erosion, while guiding the forest back to its prehistoric state of diversity, before being "loved to death" as Ravage described it.

After a destructive fire is doused, the consequences of the "burn scars" left behind can last for decades. The resulting lack of plants, fungi, and moisture leads to greater risk of erosion, flooding debris flow, and water contamination. Around the Platte, steep hills, canyons, and erosion-prone soils make for a kind of high-speed downward chute for floods and debris following a fire.[8] If, in addition to logging and planting in a manner that restores the original forest profile, they could convert the wood chips, slash, and even carbonized post-fire landscapes into biologically active, water-retaining soil, they could encourage faster recovery from catastrophic fires, and hopefully encourage conditions to prevent them from happening again.

We parked on a ridge across from a trailhead on Berrian Mountain that led to the first test site. Moments after crossing the road, we passed a tree that had been shattered and twisted by a lightning bolt, shards of wood strewn about. Walking farther, Ravage plucked wild raspberries and kinnikinnick from bushes growing along our path, handing them over for me to try. After stepping over a jeweled puffball mushroom, he plucked it from the ground and broke it open. Seeing that its interior was pure white, he deemed it safe to taste and handed me a small piece. The wan, chalky flavor was intriguing, but I discreetly spit it out as we carried on. One generally shouldn't eat uncooked mushrooms.

At last we came upon the treatment site, staked out amidst a smattered stand of lodgepoles. Loggers had thinned the area's trees some nine months prior, leaving the forest floor buried in drifts of chipped wood. "They just spewed it all over the place," Ravage grumbled. "Where the chips are more than four or five inches deep, you'll get no regeneration of plants." The remaining lodgepole might have been better left in tighter clusters, he added, as they grow in nature; with their shallow roots, they stay windfast by growing in huddles, anchoring made still more resilient by mycorrhizal fungi.

The wood chips were also at the center of the fungus tests. With all the logging going on in the area, the hope was that the chips could be quickly broken down into the makings of soil; after a catastrophic fire, inoculated chips might help to kick start the ecological recovery in an otherwise sterilized landscape.

In the test pile, wood chips had been divided into a four-square grid, representing control plots with different degrees of inoculation using native species of fungi that Ravage had cloned in his home lab. Metal stakes outlined the test plots, with signs asking they not be disturbed; the sites had been chosen in part so as to avoid interference from passing hikers, although Ravage suspected a local couple of occasionally picking the mushrooms that grew there. At first glance, the beds of wood chips looked pretty much the same. But digging under the top layer, it was evident that after five years, the inoculated plots had a much higher moisture horizon, holding more water that settled closer to the surface.

The primary tools for assessing the sites were a tape measure, compost thermometer, and camera. By comparing moisture levels and tracking the varying levels of breakdown, they could assess the degree to which fungi were digesting the wood chips. Even the control plot, with its chunkier chips and lower moisture level, had fine fibers running through it, although these may have been a ubiquitous, mycelium-like bacterium known as actinomycetes. Bacteria also play a massively important role in breaking down organic matter. But in these chipped-up trees there was not much nitrogen for them to use as food, making it a very slow process when compared with wood-rotting fungi.

"These wood chips are going to persist for decades on their own," said Ravage. "We're trying to come up with techniques that just kick-start the ecosystem. And the thing about wood-rotting mushrooms is they don't require any nutrient other than cellulose, and nothing else considers cellulose a nutrient, pretty much."

The main wood-rotting fungi are called white rot, brown rot, and soft rot, so named for the appearance of the digested lumber they leave behind. White rotters are among the few living things that can break down lignin, a durable, water-repellent polymer that's essential to circulation in trees and other vascular plants. What these fungi leave behind is pale cellulose, hence the name *white rotter*. Brown rot fungi, on the other hand, eat cellulose and leave behind lignin, often in regular cubic chunks. When your newspaper turns yellow, it's because of a reaction between sunlight and the lignin lingering in the underrefined pulp. Soft rot fungi are similar to brown rot, and the two are often confused, although soft rot typically prefer higher moisture and lower lignin content, and create unique patterns of decay.[9]

These are called primary digesters, meaning they are the first of a series of microbes that step in at various stages to break down wood, leaves, insects, and other biomass that would otherwise just pile up over time. We largely have wood-rotting fungi to thank for forests that aren't just vast tracts of dead logs, releasing their stored carbon and leaving secondary and tertiary and further decomposers to munch on what's left.[10] Over time this succession process builds soil, adding to the water-carrying capacity and facilitating seedling growth. Leveraging this process could potentially aid the return of biodiversity in scorched landscapes.

The test results underlined the point. After three years, the height of the control plots had dropped by about two inches from a twelve-inch baseline; the inoculated plots had dropped by five inches. After five years, the untouched plots dropped by three and a half inches; the inoculated plots by nearly eight. Interestingly, over the first year, the height of the inoculated plots actually increased slightly before dropping down again, which Ravage chalked up to the initial absorption of water into the mycelia, causing them to swell.

The composition of the chips revealed even more. Ideally, most of the chips would become entangled with mycelia and then be broken down into compost. By the second year, the test piles were mostly myceliated, with a thick layer of untouched wood chips atop them, and a thin layer of compost. By year five, the ratio had flipped, and the inoculated piles were almost entirely compost, save for a thin layer of undigested chips on top. Proving the point further, CUSP conducted what's called a friability test; by taking a liter of chip material at the beginning of the five-year period, and then again at the end of five years, and comparing how much material passed through a wire screen, they could measure the degree to which the wood chips had decomposed. On the first day, only about 18 percent of the as yet undigested chips made it through the screen. By the end of the five-year span, 26 percent of the control plot's uninoculated chips passed through, indicating some degradation, perhaps by local fungi, bacteria, or weathering. By comparison, when testing the inoculated pile, 73 percent of the material passed through.[11] The implications strongly suggested their plan could really work.

"If we can take those dead trees and chip them, and inoculate it with mushrooms and turn it into soil, into the organic component, into compost, then it will attract and give a bed for all the bacteria and all the mycorrhizals," Ravage said. But at the scale of the burns like the Hayman Fire, you're looking at a lot of fungus in a lot of wood chips. "It can potentially kick-start the regeneration of wild landscapes, but we've got to be able to do it at the megaton level."

The second site we visited was located on private land, owned by a couple of doctors, on a hillside that looked out onto a grand valley. Up a rugged road fenced in by a big metal ranch gate, two broad, rectangular piles of wood chips had been laid out, ninety feet long by fifty feet wide, and about five feet deep. Tucked up against the tree line, these piles were much larger than the previous test. One had been inoculated with oyster mushrooms, the other left alone. I peeled my eyes away from the spellbinding mountain valley to compare the two plots. Over the course of three years—this section of the test had started later than the previous one—the test plot had shrunk nearly five feet. Over that same time, the control plot had only shrunk by about a foot. Again, the chips in the treated plot had a higher moisture horizon, and were breaking down into smaller pieces just beneath the top layer.

"The mushrooms will never get to the top layer," said Ravage as he paced out the plot, eyes scanning the surface of the pile. Moisture had seeped in, but evaporated from the surface, leaving dry wood chips unfit for fungal consumption. Just below that layer, though, they thrived. Brushing aside a layer of chips revealed a firepit-sized cluster of firm oyster mushrooms, like hidden treasure. The prolific fruiting of fleshy, beige caps was surrounded by a ring of smaller ones, all fed by the gradually shrinking shreds of wood upon which we stood. The hills were populated by elk and bears, and Ravage was relieved that they'd left a few mushrooms behind. He stooped down and gave them a hearty sniff before yanking the clusters up from their woody bed.

The inoculation process itself had been straightforward: unbagged spawn blocks were buried in the pile, set in rows at varying intervals to measure any spacing ideal for quick and thorough inoculation. Hyphae had grown out and linked into a mycelial mat that reached throughout the pile. After five years, there wasn't much left to monitor except the rate at which the piles were shrinking.

It had started to sink in just how big a difference this simple approach could make to the health of the forest. Logging is an intrinsic part of modern forest management, and the question remains how to do it in ways that are least disruptive to the ecosystem, and ideally beneficial. Part of that involves cutting and planting according to the natural mosaic patterns of healthy, fire-resilient forest that Ravage described. Another is to ensure that soils remain moist and conducive to biodiversity that can keep their ecology active, resilient, and primed to rebound from damage. Leveraging the capacity of fungi to retain water, to unlock and distribute nutrients for other organisms, and to stave off erosion all sounded great in theory. But to hold the wet, crumbling, myceliated results in my hands made it all feel real.

Back in the truck, we set off to the lab with the oyster mushrooms that Ravage pulled from the test plot. Having demonstrated their eagerness to eat and reproduce, he wanted to clone the mushrooms, and compare them with his own strains; they would likely be the same strain that was used to inoculate the piles, but they could also be something new, perhaps slightly more prolific or fast-working. Either way, they would be added to the library of local varieties Ravage had begun banking for future tests out of a lab space in the bottom floor of his home in nearby Bailey.

We arrived at his house to a dog barking from the upstairs. "Shh," he chided, as he led us to the basement door. A cat scooted past and we stepped into a dark room; the smell and sound of bubbling instantly triggering the memory of my old neighborhood pet store. In the dim light cast by a single window, it soon became clear that cats and dogs were hardly the most noteworthy organisms around. A nearly featherless parrot eyed me suspiciously as the cat retreated upstairs.[12] I completely overlooked the mushroom lab, transfixed by a glowing corridor of glass tanks filled with dark green rain forest foliage. Peering inside, I was stunned to see a neon yellow frog staring back. Ravage's penchant for cultivating life, it seemed, didn't stop at fungi; he also reared a wide variety of (mostly) poisonous frogs from South and Central America.

"These are *Phyllobates terribilis*," he intoned, indulging in the opportunity to show off his exotic amphibians. "The most toxic vertebrates on the planet. Each one of these carries enough toxin to kill ten adult men." Ravage sourced the frogs from researchers, then bred and sold them to the pet trade so that others wouldn't have reason to smuggle them in. Functionally, their vivid coloring might also not be all that different from that of some mushrooms, a form of what's called aposematic coloration. "The color advertises the fact that, if you eat me, you're toast."

Next to the door stood a makeshift flow hood, a fish tank destined to be a fruiting chamber, and a Fisher lab refrigerator that had been converted into a culture library. A COLORADO MYCOFLORA PROJECT sticker had been slapped onto its glass, behind which stacks of petri dishes and jars of spawn sat at various stages of growth, permanent marker distinguishing the species. There were aspen oysters and cottonwood oysters; *Trametes versicolor*, or turkey tail; *Onnia tomentosa*, or the woolly velvet polypore; and *Connopus acervatus*, which has no common name— Ravage had cloned it from a specimen he saw at a meeting of the local mycological club.

The strains were deliberately sourced from the local area. Ideally, the fungi CUSP deployed would come from within thirty miles of a test site; no more than one hundred. It was a fairly arbitrary number, but it reflected the philosophy of working with nature, rather than trying to modify it. "You don't want to be releasing virulent organisms into the ecosystem without knowing what they're going to do," said Ravage. "That's why we use native wood rotters, because we're not releasing invasive species, we're releasing something that belongs in the environment."

Some of the dishes had green splotches amid their white fuzzy fungi, perhaps contamination by *Aspergillus* or *Trichoderma* molds. "This, I'm almost thinking, might beat it, because the entire outer ring is actually mycelia," Ravage noted, as though assessing a slow-motion boxing match. In another dish, the fungi had actually begun fruiting underneath the plastic lid. A beat passed as I looked up and took in the vast diversity of life in the crowded basement. "This is my office," said Ravage with dry dramatic flair. The parrot chimed in with a screech.

Our visit concluded at the third and final CUSP test plot. This one was located outside the mountains south of Denver, at a postcard-perfect demonstration farm run by the Denver Botanic Gardens. The level of nearby Chatfield Reservoir had recently been raised by about a dozen feet, which meant cutting down a couple hundred acres of cottonwood trees. Cottonwood has very little economic value, subpar both as lumber and as firewood; most of it would be ground down, painted, and sold as mulch for walkways and garden edgings.

CUSP took the opportunity to instead try their lumber inoculation methods at a larger scale. The economically unviable cottonwood was mixed with conifer from a nearby sawmill and laid out in three-hundred-foot, waist-high windrows. Small flags zigzagged across the pile, indicating where spawn bags had been planted throughout at intervals of three feet. The ratio came to roughly 1 part spawn to 20 parts wood. That was more spawn than Ravage could ever hope to produce in his basement lab. Instead, CUSP had partnered with Mile High Fungi—a nearby mushroom business that started in a pair of shipping containers in suburban Denver but had since moved to a purpose-built facility in nearby Conifer. Its funky, lofted space featured a well-endowed lab with eight feet of laminar flow–protected work surface and a massive autoclave, so that spawn of the fungi for the wood-chip project could be produced quickly.

Ideally, with properly tuned and trained strains, the ratio of spawn to wood could extend as far as 100 to 1, making the process far less expensive and labor intensive than cultivating for food production, which tends to hover around 10 to 1. The wider ratio was feasible because the goal was not to produce lots of pretty mushrooms as quickly as possible, but to encourage natural processes that work much more gradually. Looking to the future, Ravage envisioned a distributed network of production facilities that could generate locally sourced spawn to create vast windrows made from trees chipped near where they were felled, perhaps alongside the access roads used to get there. Gravity, rain, and elk would distribute the resulting compost downhill.

The philosophy of using native fungi was a matter of avoiding needless modification of an environment, when a native species will do the trick. It was also a matter of encouraging levels of species diversity and biological activity that more closely mimic what nature will do anyway. "We can't try to re-create a forest the same way we build a circuit board," said Ravage. "There is no standardization in nature, and we're going to be far more successful if we promote heterogeneity."

Rot or Not

When it comes to digesting logs for the sake of healthier forests, turning to wood-rotting mushrooms makes intuitive sense. But fungi are a powerful ally in dealing with a variety of substances, including many soil contaminants generated by industry and other human activity. Mycoremediation, a subset of the broader field of bioremediation, is one of the hottest and most hyped topics around fungi, with grand claims about their ability to clean up society's messes. But as with anything, the details are usually more complicated than the claims.

In bioremediation, bacteria reign supreme among a range of biological processes used to reduce the concentration, toxicity, or mobility of chemicals and contaminants in soil and water. Enzymatic magicians that they are, fungi are often able to facilitate the breakdown of a vast range of chemicals, compounds, and other pollutants, including from heavy industry. Yet even with all the countless fungi available to work with, the field is largely stuck on wood rotters like oyster (*Pleurotus*) and turkey tail (*Trametes*) species.

"It's generally thought that white rot fungi are the bioremediation powerhouses, but that is not true," said Dr. Lauren Czaplicki, a young environmental engineer specializing in contamination-busting soil fungi. We met for coffee near Boulder, piercings glittering as we adjusted our chairs to avoid the harsh afternoon sun. For her doctoral thesis at Duke University's Department of Civil and Environmental Engineering, Czaplicki investigated methods for identifying and leveraging the myriad wild and oft-ignored fungal communities that could become allies in restoring contaminated soils.[13]

"Fungi have been largely ignored by the literature," Czaplicki said, "From across the kingdom—the chytrids, the typical soil fungi, and the ascomycetes—*they* can also degrade pollutants. It's just that there are so many pollutants that we haven't studied all of them. So many pollutants and so many different fungi."

In the 1980s, white rot mushrooms emerged as the superstar fungus of bioremediation, after more than a thousand species were found along the Eastern Seaboard, growing from some 246 poles that had been treated with fungi-blocking creosote.[14] Finding mushrooms growing from creosote is something like finding flowers growing from a bucket of herbicide, so it caught researchers' attention.

Subsequent research was promising. Along with trees, of course, wood rotters like those in the *Pleurotus* genus were found to produce suites of enzymes that could break down hydrocarbons,[15] and remove bacteria such as *E. coli*.[16] If the toxic, water-resistant polycyclic aromatic hydrocarbons (PAHs) in creosote didn't stop them, researchers thought, they might be especially effective for detoxifying soils. Other studies showed that the same shotgun blasts of enzymes that allowed them to dissolve lignin and cellulose could degrade not only creosote but also pesticides, munitions, and chlorinated solvents, all while improving soil structure. As remediators, it was possible that fungi offered special advantages over bacteria, too, thanks to their ability to interlink hyphae and channel oxygen, or to relocate by mushrooming and tossing out spores when conditions got rough. What's more, fungi could help stimulate and even transport pollution-degrading bacteria, setting up the possibility of a kind of successionary process for decontamination.[17]

The promise of wood rotters led to all kinds of experiments to find ways of applying the fungi to contaminated soils. Usually that involved mixing it in with a wood-based substrate so that the *Pleurotus* and *Trametes*—which live on lumber, after all, not soil—could break down pollutants as a by-product of their normal lifestyle. Basically, the hope was that the fungi were messy enough eaters that their "saliva" would land on nearby problematic chemical sand compounds. The results were mixed, and with a lack of funding opportunities, the research stalled.

"Environmental engineers got really excited about it, and then super disappointed, and never wanted to look at it again," said Czaplicki. "I'm speaking in hyperbole but that's the feeling I got from my literature reviews."

Even so, the impression left by wood rotters on remediation research seemed to have lingered. "At the beginning of my PhD, I would bring my cultures over to the mycology lab and have them help me identify which ones are white rotters, and time after time they'd be, like, no that's not a white rotter you don't want that," Czaplicki told me. "I threw a lot of cultures away before I did the survey of the fungal tree of life, and found that not just wood-rotting fungi could break down pollutants."

Czaplicki has become an advocate of what is sometimes called "biostimulation." That is, rather than bringing in an outside organism to do the work you want done in a toxified landscape, you find ways of activating the latent, native organisms already mixed in with the con-taminated soil. If Ravage was practicing local species and strain selection to inoculate wood chips, this approach was *hyper*local.

When decontaminating a site, there are understandable pressures to work quickly and thoroughly, and with trusted methods, even if those methods aren't the most effective or sustainable. In the soil remediation industry, the most trusted methods are either bacterial or mechanical. "If it's groundwater, they'll pump it out, treat it, and put it back in. Or if it's soil, they'll just dig it up and move it somewhere, because they know that works," Czaplicki explained with evident dismay. "There are a lot of pollutants that bacteria can't deal with, and if the alternative is digging it up and removing it to some landfill, that's not sustainable, and it's not addressing the pollution."

The question that Czaplicki and her five collaborators wanted to explore was whether there was a way to look through DNA to identify the native fungi living in contaminated soils, specifically any overlooked species that might be targeted (that is, fed) and activated as remediators. If they could survive in those toxic environments, it seemed a good bet that some of them could also break down the contaminants. Or, maybe more interesting, stimulate an orchestra of surrounding microbial communities to do so.

"The tack I took was to see what fungi were already established in the soils, and then use those to grow and degrade the pollution," Czaplicki explained. To explore this concept, she and her colleagues visited Atlantic Wood Industries, a Superfund site right next to the Elizabeth River in Portsmouth, Virginia. The fifty-acre site, along with thirty acres of sediments in the nearby river, was rich with PAHs, along with other contaminants. In particular, a wood-treating facility that operated there from 1926 to 1992 used and stored the incredibly tough PAHs, creosote and pentachlorophenol. For a time, the US Navy leased part of the property as a waste disposal site, leading to contamination by heavy metals such as copper, lead, zinc, and arsenic, among other forms of toxic sludge, much of it easily visible to the naked eye. It was one of the most contaminated sites in the country.[18]

The EPA had dug up piles of contaminated dirt that the team of researchers then sampled along with a pile of wood chips, visually sorting them from most to least contaminated. The soils were the most contaminated, in some cases as much as 1 part PAH per 1,000 parts soil. The wood chips, deceptively, looked to have the lowest level of contamination, but still managed to stain scientists' gloves.

The samples were sent off to a lab for so-called next-generation DNA sequencing, which essentially meant they could identify every genera of fungus in the contaminated soil and wood chips. As a legacy of wood rotter fever, it is basidiomycete fungi that typically top the list of go-to remediators. But research had shown that fungi in the Ascomycota (think yeast and mold), Chytridiomycota (think mold), and other phyla had shown promise. Comparing the results to a library of fungi known to be involved in degrading the contaminants, they could see what other potential allies might be lurking in the piles. The results were impressive. Of the most contaminated samples, 60 percent of the fungi in one and 30 percent of fungi in the other were known to breakdown PAHs and other contaminants. Of the least contaminated piles, the "candidate" fungi added up to 20 percent in one, and, astonishingly, 100 percent of the last, although that was probably just a result of less fungal diversity in the sample; if there are just two species of fungus in the whole pile and both are candidates, that makes for an impressive-sounding 100 percent.

All this pointed to the fact that relatively new DNA sequencing techniques could reveal previously undetected diversities of fungal remediators in contaminated sites. It was a broad-stroke beginning that could lead to more precise and effective methods down the line. Knowing that, say, cellulose-eating fungus produces enzymes that will help crack open a particular toxic compound, you might try feeding those fungi by introducing ground-up leaf litter to the contaminated soil. It's like shouting "Where are all the ascomycetes in the house?" and instead of shouting back, they produce chemicals that detoxify the soil. (Some fungal metaphors work better than others.)

"What this study told us was that the processes involved in eating the cellulose and chitin were different, but the *communities* of fungi and bacteria are both effective," Czaplicki said. "In both cases, they weren't just eating the chitin and cellulose, they were putting enzymes out into the environment that transformed the pollutants into food that bacteria could eat."

These are promising first steps toward a more precise, less disruptive or costly way of leveraging microbes to clean up soil contamination. But as with all new and unproven methods, it is difficult to attract money and interest. When we spoke, Czaplicki seemed dispirited about the chances of securing the necessary support to move these methods forward, and to realize the remediating potential of the fungi beneath our feet.

Having grown up in Toledo, Ohio, Czaplicki had firsthand experience with the legacy of contamination. The city was once a center of industry, producing large volumes of automotive glass for nearby Detroit. Situated on the Maumee River, crisscrossed by highways, it eventually became littered with abandoned coking plants, glass manufacturing, and other heavy industry, soaked in PAHs, PCBs, and other recalcitrant contaminants.[19]

She also saw a correlation between income and environmental health. "The surrounding area isn't zoned for residential use," she said. "Those zoning restrictions end somewhere, and where they end is where the majority of poor people live.[20]

"I was endlessly frustrated by the knowledge that these companies could get away with it, and that there was no real sustainable way to clean it up," she said. "But mostly it just seemed so unjust. So I went to Ohio State, became an environmental engineer, learned all of the physical chemical treatment processes. And then I came to the realization that all these really cool ways of cleaning up soils were not getting used, because they were too expensive for these sites that were very low priority, that were also around really poor people."

My questions about Czaplicki's work with fungi seemed to rekindle a repressed excitement for a subject she had put to the side for reasons of feasibility. Funding had not been forthcoming for the kinds of remediation projects she hoped to advance; institutional interest was scarce; the prospects for realizing the true potential of her findings dim. "A lot of the resistance that I've experienced to adopting more innovative remediation approaches is that they're not proven, so no one really wants to have their site be the first one," she said.

Gradually, a conviction that it was possible to "save the world, one tract of land at a time" had given way to a sense of resignation about the intractable challenge of environmental injustice. Without money, standing, or advocates for trying new techniques in positions of influence, the status quo seemed secure. In response, her focus had shifted from one of developing new remediation methods to launching a consultancy with the goal of supporting educational efforts in affected communities, emphasizing biostimulation; rather than stimulate the soil directly, she hopes to stimulate the people connected to it, a focus on the social level rather than the soil level.

"You've just got to give them the tools they need to do it," she said as we finished our coffees. "The way to get the people in these communities, who have, like, three jobs, to pay attention, is to make it fun, to make it feel good to engage in them. So it doesn't have to be all doom and gloom: 'Don't eat this, don't let your kids play around this, don't do that.' It could be, 'Let's start a community garden,' and you get them there and you start talking to them about microbes. And then they have fresh food, and some interest. It's kind of a slower process, but it's got to be more emergent than somebody going in and being like, 'Hey, I know the solution to all your problems, let *me* do this.'"

Underground Movements

A summer sunset withdrew to a vaulted night's sky as visitors trickled into the Jean Cocteau Cinema in Santa Fe. It was a full house at the funky, one-screen theater on Montezuma Avenue, but as folks settled into their seats with sodas and spiced popcorn, it wasn't for a movie. Indigenous activists and their allies were gathering to discuss the history and future of a movement for self-determination. Outside, the marquee read: THREE CENTURIES OF PUEBLO RESISTANCE.

Seven women from the Pueblo communities of northern and central New Mexico (or womxn and Pueblx, in the event's preferred gender-neutral vernacular), all but one of them under thirty years old, took turns speaking to the history of anti-colonial struggle in and around the ancestral Tewa lands where we sat. Setting the tone for the evening, each speaker read a section of the newly drafted manifesto of the Pueblo/a/x Feminist Caucus of the Red Nation, an Indigenous-led, anti-colonial organization that helped organize the event. They weren't mincing words. "For our culture, traditions, languages, and lands to continue to exist, hetero-patriarchy, white supremacy, settler-colonial violence, resource extraction, and capitalism must die," read Nicole Martin of Laguna Pueblo.[21]

Over the course of the evening, the room grew charged with righteous resolve to overturn an old and untenable status quo. Seated behind a hand-painted sign that read in bright letters, THIS IS PUEBLO LAND, the panelists represented a modern alliance of several Indigenous-led organizations.

Serving as context and inspiration for this conversation was the Pueblo Revolt of 1680, during which the Tewa spiritual leader known as Popé organized an Indigenous alliance that forcefully ousted Spanish colonizers, a period of reclaimed sovereignty that lasted for twelve years.[22] Activists described the subsequent conquest of the region by Diego de Vargas in 1692 as the beginning of a continuous period of colonial occupation, shifting from Spanish to Anglo settlers and ultimately the United States government and the military-industrial complex.

The three-hour discussion touched on a range of issues facing Puebloan communities, many of which will be familiar to Indigenous tribes and nations throughout the continent: the use of blood quantum as a test of identity and barrier to the exercise of sovereignty; the personal, social, and civic challenges of integrating with colonial institutions while carrying forward with traditional ways of life; the lack of resources, economic and otherwise, available in the places where Indigenous people live and work; the violence, in all its various forms, that affects Indigenous people, and women in particular, at disproportionate rates.[23]

Another common factor is environmental racism, a term that gained popular parlance after a 1987 report by the United Church of Christ Commission for Racial Justice, which investigated the government's pesky habit of situating toxic waste sites and other contaminating facilities or processes near communities of color. Among the report's findings was the striking figure showing that "approximately half of all Asian/Pacific Islanders and American Indians lived in communities with uncontrolled toxic waste sites."[24] In New Mexico, the ongoing legacy of environmental injustice means that Native Americans experience higher rates of kidney, liver, stomach, colorectal, and cervical cancers than any other racial or ethnic groups.[25]

To many in the communities around Santa Fe and the northern Rio Grande, the looming, resonant symbol of state-sponsored contamination is Los Alamos National Laboratories, where the atomic bomb was designed and built. The lab sits high atop the Pajarito Plateau, overlooking the Pueblos and towns north of Santa Fe, surrounded by more than a dozen tribal nations in a fifty-mile radius. It exemplifies what the "Pueblo Resistance" panel characterized as nuclear colonialism, against which fungi have been enlisted as unlikely allies.

———

Follow the ribbon of highway that leads into the mountains northwest of Santa Fe, and after about forty minutes you'll arrive in Los Alamos. The picturesque desert suburb is WHERE DISCOVERIES ARE MADE! according to the stone slab welcome sign. On the eastern slope of the Jemez Mountain range, it's a landscape etched by dry winds and summer rainstorms that sweep over a range of fingerlike mesas, with steep canyons that drive eastward and down toward the Rio Grande river below. Just south across Los Alamos Canyon, the facilities of Los Alamos National Lab occupies a thirty-six-square-mile stretch of wilderness peppered with sagebrush, piñon pine, juniper, and the occasional hazard signs marked with nuclear contamination symbols.

Established in 1943, LANL was the epicenter of atomic weapons research throughout the Manhattan Project. Known by various code names but often simply referred to as "the Hill," it is where "the Gadget" was designed and assembled, a cutesy nickname given to one of the most terrifying technologies ever devised. In the predawn hours of July 16, 1945, about two hundred miles south of Los Alamos on the horizon-spanning flats of the Alamogordo Bombing Range, the first nuclear mushroom cloud arose with a monumental bang, blinding light, and grave portent.

In the period after the development of the A-bomb, through the Cold War and up to the present day, LANL has remained a highly active site of secretive weapons research and testing. Along with that activity have been the inevitable spills and leaks that come with trying to store or dispose of radioactive and otherwise toxic or hazardous materials. Since 1989, LANL says it has reduced the 2,100 "potential release sites" by some 40 percent.[26] Assurances of progress in cleaning up these legacy sites has done little to assuage locals' concerns about lingering and ongoing contamination.

A resonant example is the sixty-three-acre disposal site known as Area G, the construction of which began with the demolition of five Pueblo ancestral sites.[27] Reports say that thousands of cubic feet of contaminated waste sit among thirty-five pits and two hundred shafts on land considered sacred to many Pueblo peoples, less than twenty-five miles from Santa Fe's main plaza.[28] Radioactive particles are of obvious concern, but so are a variety of other compounds and heavy metals related to weapons tests, power generation, and day-to-day operations, with disposal methods that have included open detonation and burning of hazardous waste.[29] The documented and potential emissions make for a long list that includes perchlorate, chromium, and the explosive chemical RDX.[30] Having begun life as a largely unregulated, top secret compound, the lab is criticized for taking little responsibility for the consequences of its activities on environmental, population, and worker health.[31]

In 2008, at her home in Santa Clara Pueblo, about fifteen miles from LANL, Beata Tsosie-Peña began noticing the percussive sound of distant explosions. "I would come to learn later they were bombs that had been classified as too dangerous to transport for disposal," she told me. "The way they dispose of them is by open-air detonation on top of our mesas." In this monsoon-prone, windswept part of New Mexico, she had no doubt that the fallout from those explosions would find their way into the surrounding communities' gardens and waterways.

Tsosie-Peña is the Environmental Health and Justice program coordinator for Tewa Women United. Since its founding in 1989, TWU has grown into an influential community organization, with programs aimed at providing a wide range of material and social resources for local women. Originally from Santa Clara Pueblo, one of the two that directly borders LANL, Tsosie-Peña is an outspoken critic of the lab, and a passionate advocate for reviving traditional Tewa concepts and practices of land stewardship, which includes leading a variety of community agriculture and dry farming projects in the nearby town of Española. Still in her early forties, Tsosie-Peña said she remembered a period when numerous elders mysteriously died of cancer. "Everybody says that we didn't have cancer before the labs came here."

LANL and other nuclear weapons facilities throughout the country, including nearby Sandia Labs outside Albuquerque, have become notorious for dragging their feet or outright denying compensation to employees who report radiation-related sickness after working there.[32] Activists and community groups describe decades-long struggles for meaningful public participation in LANL's decision-making process, particularly with regard to the cleanup of legacy waste sites. Frustrated by what is characterized as a byzantine bureaucracy bent on self-exoneration, for Pueblo communities it represents a link between the modern issues of contamination and centuries of colonialism.

"You'll see this anywhere across the country, with any toxic site where these industries are going to be set up next to poor communities of color, and next to Native communities because they are then less likely to protest any kind of job creation that happens in these communities," Tsosie-Peña said at the "Pueblo Resistance" panel. "So then it becomes used as a weaponized tool for economics, to where our state is dependent on the war economy for any kind of ability to have a living wage for our peoples. And that's injustice."

In 2017, Tsosie-Peña was approached by artist and amateur mycologist Kaitlin Bryson. At the time, Bryson was an MFA student in Art and Ecology at the University of New Mexico, studying art and mycology concurrently, and working on a graduate thesis focused on the environmental justice issues associated with LANL. Her artworks took the form of "soft sculptures," fiber assemblages that were inoculated with fungal mycelia chosen to address specific contaminants in the soil—especially of *Pleurotus* and *Trametes* for their demonstrated remediation potential. The works were intended ultimately as offerings, gestures of respect and cleansing for the land where they were buried. Bryson hoped to make these offerings in Española, in collaboration with TWU and the surrounding community, at a pair of garden sites that Tsosie-Peña created and oversaw. At the Española Farmers Market community garden on Railroad Avenue, toxins such as RDX, hexavalent chromium, and arsenic had been detected at levels exceeding EPA standards.[33] Down the road at the Healing Foods Oasis, a permaculture food forest established on an erosion-prone slope beneath a municipal parking lot, there were concerns about petroleum by-product runoff.

Tsosie-Peña was enamored by the mycoremediation idea. "She said, 'Oh we've been looking for something like this!'" Bryson recalled over the phone. In partnership with TWU, the project quickly moved forward, and she buried her mycelial offerings in ceremonies at various sites around Española. She also led free community workshops to teach the basics of DIY mycology, and mycoremediation. At one workshop, oyster mushroom blocks were buried in the dry soil of the two gardens managed by Tsosie-Peña; the actual remediation potential of the blocks was unknown, and likely quite minimal, but that didn't make them meaningless.

"In places like the Tewa homelands, and places of systemic environmental racism, the act of acknowledgment is incredibly important," Bryson said. "The long absence of accountability from the labs has caused such a profound violence. Through these gestures of acknowledgment, we were hoping to open doors for healing to begin." What's more, the gestures pointed to a possible practical solution to a bigger problem creeping toward Española.

From 1956 to 1972, workers at LANL regularly dumped water laced with hexavalent chromium from the cooling towers of a steam power plant into nearby Sandia Canyon.[34] Chromium 6, or hexavalent chromium, is a compound involved in manufacturing stainless steel, and was used as a corrosion inhibitor in the LANL steam plant. In its soluble form, hexavalent chromium is known to be highly toxic and carcinogenic. This story might ring a bell. Erin Brockovich, and the $330 million lawsuit she spearheaded against Pacific Gas & Electric in the early 1990s, was centered on hexavalent chromium contamination of the air and ground water around Hinkley, California, and the health problems in the town that were associated with it.[35]

In 2005, LANL reported an underground "plume" containing some two thousand kilograms of the stuff creeping from the lab into the area's sole-source aquifer.[36] In 2017, the US Department of Energy, which owns LANL, awarded a ten-year cleanup contract worth $1.39 billion to a consortium called N3B. Among the biggest contamination concerns that N3B was tasked with addressing was the chromium plume. On a map, the plume looks like a mile-wide blob stretching east from Los Alamos, flattening out right along the border of the sovereign nation of the Pueblo de San Ildefonso.[37]

According to the local advocacy group Concerned Citizens for Nuclear Safety, LANL detected the plume almost two years before reporting it.[38] Since then, chromium levels at the center of the plume have read upward of 1,000 PPB (parts per billion).[39] New Mexico's standard was 50 PPB; the Environmental Protection Agency's standard is 100 PPB.[40]

As it crept toward a major source of drinking water, the DOE and the NMED made various efforts at slowing its rate of expansion or reducing its levels of concentration. That included the expensive process of pumping out the contaminated water from nearly a thousand feet below the surface, processing it to remove the chromium, then pumping it back into the aquifer.[41] Like a movie monster though, the plume kept creeping along, and locals worried about the health of their lands and water.

In 2018, a cleanup permit related to LANL's chromium plume was up for renewal; it had originally been issued without a public hearing, but pressure from local community groups managed to secure one through what was described to me as a grinding bureaucratic process. Tsosie-Peña and others in her community saw an opportunity. Communities for Clean Water, of which TWU is a core member, along with a coalition of community members and activists, including Bryson, organized a strong showing. In part, it was to express their concerns about the specifics of the permit, which included pumping and spraying as much as 350,000 gallons of treated water a day into open, lined pits, across various sections of landscape and on dirt roads.[42] Members of the community worried that the process would contaminate the topsoil, which would then blow into neighboring agricultural fields, or find its way back into the water supply. But beyond challenging the details of the permit itself, the hearing also offered a chance for the community to insist on being part of the decision-making process; the argument they intended to make was for mycoremediation.

"We know that our natural systems hold the key for collaborative systems in our communities," Tsosie-Peña told the "Pueblo Resistance" panel. "We're using myco- and bioremediation and restorative, natural systems that can clean up these waste sites to zero parts per billion. We need to all be advocating that that happens on a large scale."

On November 7, 2018, nearly one hundred people filed into the courthouse on Trinity Drive in Los Alamos, named for the site where the first "gadget" was exploded. Tsosie-Peña was among the first to speak. "Throughout my years learning of the environmental impacts taking place at LANL, my deepest fear has been for our waters," she began. "It has been so hurtful to see how our rights as Indigenous people have not been honored, how so many of our requests are not taken seriously, or are ignored or marginalized. It is vital that meaningful relationships are established with all of us, that all of our collective expertise and insights are treated equally and with just consideration.[43]

"I am in favor of the mycoremediation strategies that will be presented today, and hope that a pilot project will be required to further study these methodologies. There are people already to help with this. I also ask that the Tewa peoples be given access to go to this permit site to be able to use our spiritual strengths for water with reasonable privacy several times throughout the year, and as an integral part of any bioremediation strategies that will be enacted."

Emily Arasim, twenty-four, a CCW representative from nearby Tesuque, took the podium to speak on behalf of the local youth. "Ultimately, as young people speaking now for generations to come, we know that the only real way to heal the harm done by LANL is to completely stop all nuclear and toxic production," she said. "However, in the interim period, we support proposals to address chromium contamination, such as mycoremediation, which has the potential to be a community-safe, ecologically sound, and truly restorative way to address the massive harms done to our communities and lands through chromium contamination."

Among those invited to testify about the efficacy of mycelium for cleansing the contaminated plume was Peter McCoy of Radical Mycology, at the invitation of Bryson and with funding support from CCW and others. He laid out a case based on his review of research into the ability of *Pleurotus* mushrooms to capture hexavalent chromium. The approach he pictured involved spent substrates from mushroom cultivation put to use in filtering chromium-laced waters. These materials might be supplied by local mushroom farms, he said, encouraging the growth of a new sector in the local economy. If no such businesses already exist, he suggested, the skills were easy enough to teach, a fact to which he could attest. He explained the ways in which these mycoremediation methods were actually superior to bacterial bioremediation, which LANL had already tried. For instance, while the latter breaks chromium 6 into the less harmful chromium 3, those molecules can still potentially revert to their former, toxic form.

"Over a short period of time, sometimes up to an hour, one hundred percent of the metal will stick to the mycelium," McCoy told the hearing. "The tissue can then be removed and actually washed off, and [the metals] used in industry, recycled. Effectively, you are completely cleaning the water of the chromium, removing it, not just trying to transform it, or even filter it."

For nearly five hours before a technical portion of the hearing began, speaker after speaker articulated their frustration at the seeming indifference of the labs with regard to the wishes of their community, a dynamic they hoped to change through a collaborative mycoremediation project. "If granted, this project would be a collaboration between myself, Mr. McCoy, First-Nations student interns from surrounding Universities, and LANL," Bryson explained. "We believe that it is vital that the contamination issues be brought to the forefront of our culture and addressed collectively. It is imperative that we all have a voice in this matter, and that we are all allowed to be invested in the process of remediation. In this way we can form a partnership that fosters growth rather than continues to isolate and fracture."

Here was a collective appeal to the neighborhood nuclear power to honor the agency and will of the people on whose land they operated, and which they had contaminated with toxic waste. Rather than fight, the community proposed collaboration; it was a beautiful gesture. It might not surprise you to learn how it turned out.

Just a few months after the hearing, the New Mexico Environment Department that issued the permit in question simply validated the original permit. The reasons were largely bureaucratic: limits under the given permit and the mandate to exhaust interim treatment methods. "Current water treatment methods are sufficient to meet and exceed the applicable groundwater and drinking water standards," read the report from the hearing officer.[44]

So it goes.

———

"Wow, look at that bee balm!" exclaimed Tsosie-Peña as we began our tour of the Española Healing Foods Oasis. Under the searing sunlight of a gem-blue midday sky, the mint bush's flowers looked like tiny purple fireworks suspended in mid-burst, enlivened by rains that had swept the parched valley in the five days since her last visit to the hillside garden.

The Healing Foods Oasis was situated on a slope overlooking the green lawns of Valdez Park. At its top was the parking lot for the City Office of Española, a town of ten thousand people twenty-five miles north of Santa Fe. The park was designed as a food forest, according to permaculture principles—or Indigenous sustainable design, as

Tsosie-Peña, who is certified in permaculture design, quickly corrects. "It's Indigenous knowledge, let's call it what it is," she said.

The drought-tolerant plants were thriving with the recent rainfall, most of them native to the area. In fact, it was rain that had inspired the creation of the garden in the first place. "I was out here with my kids one day during a rain event, and saw how much water was getting washed off this slope," she told me as we stepped past a burst of cheerful yellow sunflowers and over a wide mound of recently laid wood chips. A local focus group had expressed concern about environmental justice issues in the community, and among the initiatives they proposed was an expansion of community gardens in Española. It took four years just to get approvals, and after just as many years since the first plantings, the garden was thriving.

A dragonfly hid among the bristled contours of a rosemary bush. Prayer flags fluttered from the branches of a scrawny sapling, with wishes written by children at the last Regeneration Festival when it was planted; around its roots were buried pebbles into which they had imparted their negative or hurtful thoughts. Everything at the garden had been planted with intention and purpose, including several plants often categorized as invasives, like goat's head and ragweed, but which play valuable ecological roles and have many traditional uses.

Near the top of the hillside, burgundy stalks of amaranth grew in short rows, part of a collaboration with Guatemalan farmers to rematriate the plant within the landscape. Amaranth is one of Tsosie-Peña favorite plants, in part because of its fundamental role in precolonial traditional diet and culture, but also because of its tenacious resistance to being outlawed or destroyed, labeled as a "superweed" resistant even to Monsanto's glyphosate-based herbicides.[45] "Freaking *Roundup* can't kill amaranth," she said with a note of deep appreciation as we stood at the edge of their neat ranks. "It's like the plants are in the revolution."

All around us was what had once been an angular, barren slope that rainwater carved up and sloughed away every year. The scant eleven inches of annual rainfall were plenty to emulsify and move the parched dirt. The city's original response was to pay annually for heavy machinery to scrape off and move the vegetative layer, so Tsosie-Peña had pitched the oasis project as a sustainable alternative. As we walked, she noted the lack of any erosion even after the recent rains, a sign that the roots were taking hold. The plants brought life and stability to an otherwise dusty hillside, but more important, they served to reconnect members of the community with their land, and the practice of dry farming practices that had sustained Pueblo people in this arid place for millennia.

"The intent is to reconnect people with the plant relatives," Tsosie-Peña said. "It's really tapping into our cultural strengths as land-based people, that we share across cultures here in the valley, and people here are really strongly connected to that." She spoke with a tone of resignation toward the notion of reforming the institutions that had rendered the nearby lands and waters less safe. But also in her voice was a sense of quiet resolve. "When the system collapses, then, well, what's in place?" she said. "That's the knowledge we're going to look to."

At the top of the oasis, under the shade of an oak, the parking lot abuts sharply, separated from the slope by a line of telephone poles laid down end to end. Tsosie-Peña gestured to the ground on the side of the garden. "This last year we buried mycelium pillows instead of tree planting," she said, chuckling. She estimated about fifty bricks of oyster mycelia had been planted along the swale in one of Bryson's workshops.

"We had all the holes pre-dug, and all the youth took a mycelium brick and buried it in the ground," she said. "It's probably too hot for some of the mushrooms, but we hope that they're spreading underground."

CHAPTER ELEVEN

MYX MESSAGES

If you stepped into the art gallery of Hampshire College in January of 2018, it would be reasonable to think that it had come under occupation by a mysterious government agency. In the center of the echoing, cement-walled space was an austere wooden desk, neatly set with a globe, telephone, lamp, stack of newspapers, and a name placard that read COMMUNICATIONS. Documents lying about offered a clue about the occupants of this provisional office: *The Plasmodium Consortium*.

This consortium appeared to be involved in some substantial work. A map of proposed routes for public transportation in the surrounding Pioneer Valley of Western Massachusetts took up an entire wall. On another, smartly designed placards and data visualizations related research into addiction, food deserts, and other issues of public concern. Accompanying each example were copies of advisories that had been mailed to the top offices of the federal government; for instance, then-attorney general Jeff Sessions's in-box received a report on cannabis policy. The Plasmodium Consortium could be thought of as a think tank of sorts, although none of its members *thought* in any sense most are familiar with; they didn't even have brains. The findings had come from snotlike blobs in petri dishes.

"Our researchers are uniquely qualified to undertake this investigation because of their objectivity," read the consortium's policy advisories, penned by artist and experimental philosopher Jonathon Keats. The slime molds, delivered by Carolina Biological Supply, represented Hampshire College's first-ever visiting nonhuman scholars, a position that came with visiting hours and even office space, albeit not exactly a corner suite. "A dark, slightly damp basement space was kind of perfect," said Amy Halliday, who at the time was the college's chief curator, as well as the consortium's designated outreach coordinator. "Space is always at a premium at Hampshire, but luckily the slime molds have different desires."

Slime molds are easy to picture even if you've never seen one. *Physarum polycephalum*, a particularly well-studied species, takes the form of a gelatinous, bronze-colored goo. This is the plasmodium, a slithering slime that moves in glacially slow, searching waves comprised of pulsing, veinlike masses. To witness their movements requires use of microscopes, or time-lapse video, in which they look and behave exactly like the Blob. Usually just a few inches wide, but sometimes as large as two feet in diameter, they're commonly found in foamy splotches on the trunks of trees, with an appearance that is often aptly compared to dog vomit.

Despite such inauspicious associations, the more you interrogate a plasmodial slime mold, the more you understand why they've become subjects of intense interest in recent decades. Mathematicians, city planners, game designers, astrophysicists, computer engineers, researchers in all manner of fields have found deep insights and mysteries alike in slime molds. Why? For one, they are living examples of emergence, collections of simple individual parts that together exhibit complex behavior.[1]

At first glance, slime molds can be, and for a long time were, confused with fungi—they live as decentralized networks that reach toward and surround their food, producing spores from vaguely mushroomlike structures in order to reproduce. Put two genetically compatible individuals into contact, and they'll merge to form a single being. Divide and separate a slime mold, and the various parts will keep on creeping on independently.

Yet they are quite distinct from fungi. For one thing, they *move*. Rather than reaching into and absorbing their food, as fungal hyphae do, slime molds seek out and enshroud their meals within a sort of improvised stomach. Their proper name reflects a somewhat ambiguous nature: Myxomycota, or Mycetozoa, which translates literally to "fungus animal."[2] The word *polycephalum* translates to "many heads," a nod to the organism's decentralized nature, and suggestive of its ability to optimize, decide, and even remember.

Plasmodial slime molds are actually a single-celled amoeba, composed of free-floating nuclei. Yet the plasmodium can also decide to reorganize itself into millions of individual cells—from an individual to a multitude. "In writing about slime mold, one can slip between singular and plural forms at every reference with due cause," writes scholar Aimee Bahng, "as both cellular and plasmodial slime molds exist alternately as singular and plural, depending on how and when you're counting."[3]

Look closely at a slime mold, and you'll see what appear to be veins; zoom in, and more veins become visible; push in still closer, and the pattern repeats itself again, in an almost fractal manner. Like fungi, slime molds operate like living networks that constantly assess and respond to their environment, every single surface informing the actions of the whole. When they detect food, proteins similar to those in animal muscle flex the internal pathways to create flows of cytoplasm, a process called streaming, which pushes the plasmodium toward its meal and, using mechanisms still yet to be fully understood, transmit information.

In 2016, biologist Audrey Dussutour and her team at the Research Center on Animal Cognition at Université Paul Sabatier in Toulouse, France, devised an experiment to test these creatures' capacity for memory.[4] Dussotour and her team placed *Physarum* slime molds across from big cakes of oats, one of their favorite foods. The only thing standing between the blobs and the objects of their desire was a gelatinous bridge laced with either caffeine, quinine, or in the control case, benign agar. Caffeine and quinine are both compounds that slime molds find distasteful and try to avoid; they dislike caffeine in particular, which in sufficient amounts can actually harm them. As expected, the chemically obstructed slime molds were reluctant to touch the bridges, compared with the neutral control. But over the course of a couple of days, appetite won out over caffeine aversion (an aversion I cannot relate to), and the test subjects squished across to reach their oats. Given a couple of days of rest, they seemed to lose their newfound confidence, and the cycle repeated.

A follow-up test was more interesting. In it, Dussutour and her colleagues used a similar arrangement, only this time with a somewhat less distressing substance: salt. Test *Physarum* were "trained" on salt, once again developing an indifference toward the tainted bridges. But this time, instead of being given a chance to rest, they were cut into thousands of little pieces. (Don't worry, the slime mold can split and merge quite freely, given a few hours to rebuild their vasculature.) The trained fragments were then combined with the "naïve" slime molds, which had not been exposed to salt. The results showed that, when the trained and the naïve slime molds were combined into a single organism, the resulting blob retained the "learned" behavior of its trained portion.[5] In an organism without any brain, or distinct organs of any kind, such storage and transmission of information are an intriguing trick indeed.

It turns out you can teach a mindless blob new tricks. Build a maze with food at the other end, and after a bit of slithering exploration, the *Physarum* reliably and quickly finds the shortest possible route to its meal.[6] In one of the best-known experiments, a team of Japanese and English researchers, led by Toshiyuki Nakagaki at Hokkaido University, laid out a miniature map of Tokyo and thirty-six surrounding towns inside a petri dish. Slime molds aren't much interested in tourism, so the scientists represented the cities with oat flakes of various sizes. A *Physarum* was then unleashed from the center, which began stretching out to explore its surroundings in its perpetual quest for food.

At first, it expanded like a web, with large arterial channels supporting its many slender, exploratory fingers. As it found its oats and noted dead ends—marking them with pheromones, a sort of anti-breadcrumb trail—the least-utilized channels collapsed, leaving behind only those that could best direct the newfound nutrients throughout its body. The overall networks gradually compressed from slender and wide-reaching webs into refined pathways. Over the course of about a day, it had found all the flakes within the boundary of the dish, paring itself down to a lean network that looked remarkably like the Tokyo subway system, taking mere hours to mimic an optimized system of distribution that had taken human engineers years to design.[7] Similar experiments have been conducted with other metropolitan maps from around the world; the slime mold hasn't indicated its preferred city.

The ability to arrive at optimized solutions to complex problems—in the examples of the subway systems, how best to form connections among a variety of disparately spaced resources—inspired the formation of the Plasmodium Consortium. "Here was just a really explicit case of something that humans think only humans can do, because we have this incredible analytical ability, namely making freeway systems or making subway systems," said Jonathon Keats, the experimental philosopher behind the consortium. "But here's a slime mold that's doing it."

The idea of slime that can *think* is perhaps a little bit unsettling, owing in part to our total inability to understand their intentions. After all, who knows what conclusions they might come to, particularly with regard to the form-bound beings prodding them with sensors, slicing them into pieces and stuffing them into obstacle courses. On the other hand, it's cause for considering just how little we understand about the agency of other forms of life, be they animals, fungi, plants, or slime mold.

In these strangest of creatures, "We have the opportunity to reflect on ourselves, our own biases, our own problem-solving mechanisms, and our hubris," said Keats. "I'm always suspect of any sort of prioritizing of human intelligence, or really anything about us, as being different in kind or being somehow special or somehow being better than what we find manifested in other species."

The Plasmodium Consortium was an effort at putting into practice another way of thinking about thinking. By posing questions of public policy to a blob, it created a context in which we humans had another perspective to consider. The value was largely in the asking, though, since the process was entirely mediated by humans, and the answers weren't likely to inspire any changes in policy. It was an opportunity to try out methods for engaging some small part of the universe of nonhuman— and therefore, in theory, nonpartisan—agents that surround us.

"We need to recognize that if we don't give them agency, we're only going to make the situation worse for us as well," Keats said. "How do we bring them into the system such that . . . we are by some sort of consent, rather than by decree, comanaging this world in which we live, which is also lived in by all these other organisms, all of which have had agency in the past, and deserve agency?"

If this all sounds a little bit ridiculous, consider that this very reaction is what is being examined and challenged. Keats says he sees his projects as thought experiments acted out in the real world, where they can run headlong into everyday human assumptions, self-appointed decision makers for the world that we are. Of course there's nothing inherently wrong with being ridiculous, either. It might even be helpful. "There needs to be a basic level of accessibility, and there also needs to be some entertainment value, frankly," he said. "I think humor has enormous value in terms of engaging people, both because people like to laugh and also because when you laugh you're taken off guard in a way that you might be more open to surprise, or more open to ideas."

So, how do you ask a question of something as inscrutable as a slime mold? And what do you even ask?

Keats approached Amy Halliday at Hampshire College because they had both worked together previously, on a project at nearby Amherst College to build a camera with a thousand-year-long exposure time. With Halliday's help, there seemed a good chance this weird idea about slimy scholars would find interest at the school, which had been founded on the promise "to innovate and experiment, in every dimension of collegiate education where it appears promising to do so," and to "regard no cows, academic or of other breed, as sacred."[8] The school turned out to be the right partner for the thought experiment. Dr. Megan Dobro, associate professor of biology at the college, was recruited early in the consortium's formation. By folding the project into a Microscopy and Modeling course she was already teaching, Dobro effectively integrated it with the school's system. "Megan is a biologist who was willing to take this seriously, and who also believes in the seriousness of play and experimentation," said Halliday of Dobro, who was on sabbatical and unavailable to talk for this book. "The science people were worried that we were going to make them look silly."

In the spring of 2017, five students began their studies of the slime mold, and in the fall joined forces with faculty and staff to devise experiments. Over the course of a series of workshops, they reviewed newspapers to draw up a list of societal quandaries about which the plasmodium might offer some insight. Participants came from fields as far flung as economics, philosophy, ecology, poetry, art, design, biology, community organizing, public policy, animal behavior, and mathematics. Eventually they pared the list down to five subjects. One group would look into how public transportation routes in the Pioneer Valley might be remapped in order to better serve marginalized populations. To do this, they would leverage the blob's ability to optimize as a network, enticing it to spread out over a map of the area, a reimagining of the Tokyo subway experiment, but one that took local sociopolitical considerations into account. Another group investigated the problem of food deserts; to simulate the availability of nutrition and propose the best distribution of farmers markets, the experiment used different ratios of sugar to protein according to the locations of existing food outlets and the quality of nutrition they offered.

Addiction was another area of study. Slime molds, it turns out, have a strong affinity for valerian root, so one experiment modeled addiction by confronting a plasmodium with the choice of either pure root extract, or more nutritious protein-and-sugar mix. Another blended root and nutrient in different ratios, becoming more nutritious toward the rim of the dish. In the former experiment, the slime mold stuck to the root and died; in the latter, it gradually moved toward the pure nutrient, suggesting to the researchers that a "gateway drug," perhaps like cannabis, "was politically desirable."

To explore the issue of border policy, students constructed three models, each employing a pair of *Physarum* blobs to simulate the populations of two neighboring countries. "Since slime mold can be understood as both a population and a system for transporting resources (nutrients), it presents a compelling model for observing the effects of diverse border conditions," read the blurb Halliday penned to accompany the border experiment. In one model, a closed border was simulated with an impermeable plastic barrier dividing the petri dish, preventing the side stocked with sugar to access the other side, equipped with protein-rich food, and vice versa. In another, a bar of light—which shadow-lurking slime molds have evolved to avoid—simulated an intermittently enforced border.

An open border was modeled by simply spreading the resources evenly throughout the petri dish. Unsurprisingly, the open border scenario led to the healthiest, happiest slime, most especially in the border zone itself. A policy proposal was sent to then secretary of Homeland Security Kirstjen Nielsen, suggesting that open borders would lead to the maximum flourishing of life. (The consortium never received a reply.)

Something that looks like the result of a hearty sneeze might not seem a promising source of wisdom. Getting to the point of approaching slime mold, asking for insight into complex social issues is enough of a hurdle, but working out *how* to ask poses its own set of challenges. To help crystalize the possible questions that students might pose, an "Iconography of Slime Mold Behavior and Capabilities" was designed early on by Thom Long, then an associate professor of architecture and design at Hampshire College.[9] Among the twenty different capacities represented on the menu were "the ability to explore an environment and rationally decide when to stop exploring in order to exploit a resource," and "the ability to transfer learned behavior to naive slime molds by fusing with them." Each of the *Physarum*'s various abilities were assigned a symbol, according to a legend that had slime molds represented by a black circle, their activity defined by a green circle, with blue and red circles representing attractant and repellent stimuli, respectively.

These clearly enumerated capacities offered a framework for designing effective experiments. For example, sporulation or fragmentation indicated stress for the slime mold; if a *Physarum* started producing spore structures in your experiment, chances are it was expressing something like distress about its situation, which would go down as a data point. But it also helped to instill in the overall project a sense of completeness, to give the consortium institutional substance and aesthetic cohesion. Along with the icons, Long also designed letterhead, research layouts, and other visual elements to foster a unified presentation of the various experiments.

On the desk in the gallery, there were take-home boxes, made from cardboard and branded with the consortium's logo and labeled MOBILE RESEARCH UNITS. Inside each was a petri dish with slime mold inside, so that visitors could devise their own experiments. All these elements lent to the sense of a legitimate agency charged with consulting our nonhuman neighbors in matters of social concern, along with the notion that such consultations could be undertaken by anyone.

"We didn't want the gallery to look like a science conference with posters," says Halliday. "All of these visual decisions had to go into the nature of the experiments, because a lot of people do experiments, but they're just to get the data and they're not meant to visually communicate."

Keats elaborated, "We developed the aesthetic to reflect the reality that it was a policy think tank, and to bring people into that reality according to a vernacular that would already be familiar to them, in order for people to fully engage the reality and to take the output of the think tank seriously."

For all the artifice involved, the Plasmodium Consortium was not intended merely as an exercise, nor a metaphor, as it is often interpreted, according to Keats. Rather it was meant as an *enactment* of more intentional engagement with another form of life than is typical of our species or, at least, our culture. The result is a certain sense of absurdity, but that may be just what gets people to give a second thought to the attitudes and assumptions it challenges. Those attitudes have helped usher our planet into the age of the Anthropocene and mass extinction, after all, so which is the more ridiculous?

"Absurdity is infinitely deep in its own right," Keats said. "But the absurdism of the slime mold solving this problem of border walls or of drug addiction, as a first point of contact, is potentially a way in which to start to recognize as a public what goes into a model, and what the limitations are. These models are deeply problematic, they're really limited, but they're better than nothing, I think."

An undifferentiated blob can serve as something of a Rorschach test, representing almost anything we choose to see in it (or *them*). This is especially evident in our regard for "lower" forms of life, the marginalized or poorly understood beings about which socially rooted preconceptions and biases tend to run rampant.

Take the case of *Dictyostelium discoideum*, another slime mold notable for its habit of spontaneously transforming from a widespread colony of single-celled amoeboids into a multicellular "slug" that slowly crawls about looking for food. When the light, humidity, and pH levels are right, the ephemeral slug stops slithering, shoots up stalks, and disseminates spores that germinate elsewhere to form new, single-celled amoeba; it has ceased to be an *it* at all. To the observer returning to check where the slowly slithering slime once was, "they" have seemingly disappeared.

The question of how these slime molds, absent any brain or apparent decision-making center—or center of any kind—could coordinate actions this way was for a long time a mystery. But biologists had a ready explanation for what was happening: a special group of signaling cells, rather tellingly named "pacemaker" or "founder" cells, guided the organism-at-large in deciding when to divide into many smaller bits.[10]

Harvard biologist Birgit Shaffer outlined in 1962 how it could happen. Slime molds were known to circulate a compound called cyclic AMP, a substance commonly found in amoebas, plants, and humans alike, associated with memory, metabolism, gene regulation, and immune function.[11] Scientists determined that it must also be responsible for the signaling that led to the aggregation or disaggregation of the *Dictyostelium* cells. Shaffer described a model by which a special set of delegative cells could release cyclic AMP, sending it rippling through the rest of the organism, and inducing the rest of the cells to bind together.[12] It made intuitive sense to scientists, and the model was largely taken as read for some twenty years. One nagging problem with the theory, though, was that no one could actually *find* the pacemaker cells.[13]

This didn't sit right with Cornell physicist Evelyn Fox Keller and mathematician Lee Segel, who proposed a new way of understanding how *Dictyostelium* managed to divide without any apparent mechanism of guidance or coordination. They outlined a mathematical model that could explain the phenomenon, inspired by concepts articulated by the famed mathematician Alan Turing regarding morphogenesis, which describes the natural processes behind the creation of complex biological forms.[14] Turing had explored the ideas with a focus on flowers, but Keller and Segel saw their value in understanding the slime mold.[15]

Throughout the 1970s and '80s, they developed mathematical models and experimental evidence showing how the cyclic AMP flows that ran through the *Dictyostelium* and correlated with a phase shift could be triggered by any combination of individual cells, responding to their local conditions and forming into clusters of activity that then spread out and triggered a totalizing change throughout the organism. These principles could account for the apparent ability of a decentralized, brainless mass of cells to "decide" to disaggregate, no special cells required. It was an elegant and suggestive alternative to the prevailing ideas at the time, which seemed rooted in a certain set of socially accepted perspectives.

"Biologists, for the most part, showed little interest in our ideas, and despite the absence of evidence, continued to adhere to the belief that founder cells (or pacemakers) were responsible for aggregation," wrote Keller in an essay about their research and the reactions to it.[16] In another essay, she elaborated on what she thought was at play in the reluctance of scientists to acknowledge the insufficiency of their model.

"[T]he pacemaker view was embraced with a degree of enthusiasm that suggests that this question was in some sense foreclosed," she wrote. "Might it not be that prior commitments—ideological, if you will— influence not only the models that are felt to be satisfying but also the very analytic tools that are developed?"[17]

Science is upheld as the realm of the objective pursuit of knowledge and understanding. But of course, it is as shaped by individual and social subjectivity as most any other human endeavor. Perhaps, with all its tools of inquiry and mechanisms of objectivity, science offers a unique venue for confronting our own preconceptions and biases toward forms and ways of life that we don't intuitively understand.

Queered Science

Dr. Patricia Kaishian spends a lot of time peering into a microscope. That's the only way to study her chosen subjects, the Laboulbeniales, a generally neglected order of fungi that parasitize various arthropods. On beetles, cockroaches, earwigs, bees, bird lice, and innumerable others, these fungi express themselves as tiny tufts that sprout from the exoskeleton. One can sometimes spot them on ladybugs, on which they appear in clusters of teensy amber stalactites.

Unlike other, better known insect parasites, such as *Cordyceps*, Laboulbeniales aren't known to be deadly to their hosts, but they're also distinct in more interesting ways. For one, they don't take the form of a mycelium, instead emerging from cellular division straight out of the spores themselves, which stick to the bug and grow into a microscopic, frondlike thallus. The spores transmit through direct physical contact, demonstrating strict preferences for species, sex, and even specific body parts in their hosts.[18] After some 150 years of study, these fungi remain largely mysterious, and subjects of debate as to their classification, having been described early on as "doubtful ascomycetes," though their phylum at least has since been determined as Ascomycota.[19]

Kaishian took up Laboulbeniales as the subject of her dissertation in 2018, and has since described eight new species in publications cowritten with her advisor Alex Weir of the SUNY College of Environmental Science and Forestry (ESF), who also specializes in Laboulbeniales. Under the Forest Pathology and Mycology department, it was one of the few programs in the country that conferred a degree in mycology. To study her subjects, Kaishian drew from the insect libraries of SUNY-ESF, the Natural History Museum in New York, and the Smithsonian National Museum of Natural History in DC, sitting quietly at the microscope as a parade of insects passed beneath her objective, hoping to spot the peculiar fungus wherever it had been overlooked. It was grinding work; in one week, after reviewing some three thousand insect specimens, she found forty instances of Laboulbeniales.

"They're not very well studied by anyone; a lot of mycologists aren't particularly familiar with them, and that's largely because of their size," Kaishian told me. "There's this idea that there are charismatic fungi and there are noncharismatic; anything that's big, beautiful, and edible is going to get a lot more attention."

Admittedly, in general society, one is usually hard-pressed to find many people who will describe any fungi as "charismatic." Maybe that's quite telling.

Kaishian has a personal affinity for the misunderstood, marginalized organisms she studies. She credits this to having grown up steeped in nature; instead of going to preschool, Kaishian spent her childhood years in forests and swamps, splashing through mud, catching frogs and snakes. In reflecting on her early years in communion with these organisms of mixed public repute, she sees some of the precursors of her interest in mycology.

"I had pretty intense gender dysphoria as a child, and my relationship to those organisms felt kind of related because I didn't quite fit into either gendered group," she said. "I identify as a cis woman, but when I was a kid, I felt like I got along with these organisms in a way that I couldn't quite name, and I didn't have a name for my experience when I was going through it at the time. It was only now, when there's more discussion around it, that I can identify it for what it was; I just developed a kinship with certain organisms because they were also defying these categories, and were a little bit marginalized and demonized."

In 2020, Kaishian and her colleague Hasmik Djoulakian, a feminist educator and writer, published a paper called "The Science Underground: Mycology as a Queer Discipline."[20] Drawing from critical theory, as well as Kaishian's experiences as a working fungal taxonomist and as a self-identified queer woman, the paper synthesized a perspective on the marginalization of fungi, and the situation of mycology as a boundary-bending discipline rife with "queer" dynamics, even as it expresses many of the same institutional proclivities toward the patriarchal and heteronormative that define science in general. Its purpose was "to remediate our relationship with fungi and all organisms—thereby queerness—by collapsing and myceliating the emotional space between human and non-human," by exploring the "dogma of institutional [capital *S*] Science, as well as the biology, history, and methodologies of mycology through a queer theory framework."

Part of why mycology might resonate with the misfits and the marginalized is because fungi themselves have long been seen the same way, even within the field itself. As we've observed, forest pathology is the common lens through which fungi are assessed in academia.

"We're looking at the danger of fungi first and foremost; they are first and foremost a threat, and they are to be controlled and eliminated, that's the whole departmental framework, which is crazy to me," Kaishian said while presenting her work at the New Moon Mycology Summit in 2019.

Let us recall that mycology is often described as a neglected megascience.[21] Fungi had to wait until the mid-twentieth century before their distinct identity was recognized. Even with institutional recognition, culture at large has yet to catch up. There remains a general phobia toward fungi, and, it seems, a sense of iffiness about people who commit their lives to studying them. For those who come to fellowship with or fascination about fungi, the journey there tends to take place along the margins, too.

"Charismatic mycologists first introduced us to this queer world," Kaishian said. "They showed us that there was actually this other world that exists where the rules didn't quite make sense, and maybe that resonated on a personal level. It did to me, as a queer person."

Mycologists, she notes, are often dyed-in-the-wool autodidacts, gaining much of their knowledge outside the classroom, such as in gatherings of the sort we've visited in this book, as well as the various mycological societies and festivals and forays that, in Kaishian's analysis at least, challenge the centrality of scientific knowledge, as well as the funding model that justifies a study of fungi largely on the basis of their risk to timber board feet. Preserving the logging industry is, it should go without saying, not what brings all mycologists into the field.

Mycology is in many ways unique among the natural sciences, owing in part to the nature of fungi themselves. Much like slime mold, the body of a fungus tends to defy our intuition; at the same time, they are uncannily situated somewhere between animal and plant. And what is a fungal individual? If you clone a mushroom, are there now two of one, or one of each? In the realm of reproduction, too, they confound conventional conceptions. Take the example of *Schizophyllum commune*, also known as the split-gill mushroom. For fungi, mating types are the rough equivalent of sex; while some express three or four mating types, the split-gill mushroom has 23,328. Whatever the opposite of a binary can be called, this fungus seems to fit the bill; or rather, it doesn't fit any bill. Hence some like to call Fungi neither kingdom nor queendom, but *queer*dom. This queerness extends to the very ways scientists study fungi.

Out in the field, mushrooms aren't as easily documented as plants, which tend to stay put and don't often vanish over the course of a few days. Animals may wander, but they are easily seen and stay in more or less the same form throughout their lifespan. In the wild, though, fungi are often ephemeral, unpredictable, fickle. Many are essentially invisible, living out of sight underground or woven among the cells of a plant, and exhibiting lifestyles that science doesn't understand—that *no one* yet understands, except perhaps in some intuitive or traditional way. Indeed, intuition necessarily factors into the scientific study of fungi.

A botanist may establish a transect, a line laid down through an environment, along which they will observe and record the occurrence of a given organism. Finding mushrooms is, as Kaishian describes it, more a matter of taking a "timed wander." It's foraging, or scavenging: gathering. These are, incidentally, also highly gendered forms of labor.[22] Kaishian recounted scenes of field mycologists "rolling out of bed" to go on a walk in the woods, in contrast with the much more rigorous schedules of the botanists, entomologists, and other biologists. Even as part of a rigorous scientific study, finding mushrooms in their element is a largely intuitive affair, not to mention a matter of luck.

"You kind of just go outside and wander around," Kaishian said. "You might not find a single mushroom on a transect, but then if you say, 'You know, I kind of feel like they're going to be over here,' and you lean into that vibe, then you might find a beautiful valley of mushrooms." This, she added, didn't always jibe well with the more common logic of and approach to scientific documentation. To meet a living being in nature and assess it as a data point to be logged and aggregated into a spreadsheet may provide interesting and useful information, but what does that do to the relationship between scientist and subject?

"It does frighten me where we're going with our conservation prac-tices," said Kaishian, "because it's really about numbers, and there isn't much space for the romance, the love, or just your intuition that these things are important and beautiful and valuable, and we need to save them because of just who they are."

Mushrooms pop up where and when they deign to do so, quite unpredictably, most often without being seen by anyone. To study or, maybe more appropriately, to *know* them requires tools that aren't com-mon to the field biologist or found in their manuals. The appropriate tools may exist, though, in other traditions, other conceptions of science, other "ways of knowing."

Illustrating this notion, Robin Wall Kimmerer, a professor of forestry at SUNY College of Environmental Science and Forestry at Syracuse, and member of the Great Plains Potawatomi Nation, notes a Potawatomi word that won't be found in a biology textbook: *puhpowee*, which can mean "the force that causes mushrooms to push up from the earth overnight."[23] In *Braiding Sweetgrass*, her book about the relationships between science and Indigenous ways of knowing, Kimmerer notes the Potawatomi's animist language, and the difference it makes in the speaker's perception of the world around them.

"To be a hill, to be a sandy beach, to be a Saturday, all are possible verbs in a world where everything is alive. Water, land, and even a day, the language a mirror for seeing the animacy of the world, the life that pulses through all things, through pines and nuthatches and mushrooms."[24]

By contrast, in the English-speaking world, fungi tend to be characterized not as verbs, as living beings engaged in fluid processes of life, but are rather bound up in some of the least charitable adjectives. They are *gross*, *freaky*, *alien*. Even the more benign terms *weird*, *strange*, and *unusual* are not ones we might generally use to describe a being that we trust or respect.

British naturalist William Delisle Hay might be credited with coining the term *fungiphobia*, writing in the late nineteenth century that "no fad or hobby is esteemed so contemptible as that of the 'fungus-hunter,' or 'toadstool eater.'"[25] Hay was speaking of fungiphobia as a form of culturally driven ignorance. "If it were human—that is, universal—one would be inclined to set it down as an instinct, and to revere it accordingly. But it is not human—it is merely British." That is to say, it is the result of learned prejudice. English writer Sir Arthur Conan Doyle provided one of the more vivid articulations of anti-fungal sentiment.

A sickly autumn shone upon the land. Wet and rotten leaves reeked and festered under a foul haze. The fields were spotted with monstrous fungi of a size and colour never matched before—scarlet and mauve and liver and black—it was as though the sick earth had burst into foul pustules. Mildew and lichen mottled the walls and with that filthy crop, death sprang also from water soaked earth.[26]

Western scientific traditions are deeply rooted in Anglo history and culture, so it's little surprise that some of the reflexive derision toward the fungal kingdom (another term loaded with English context) made the trip overseas, too. Anglo aversion and disdain toward these organisms has seemingly colonized, one might say, our perceptions and conceptions of the natural world, even those of scientists, whose only lens is meant to be the dispassionate view from the objective of a microscope.

It seems safe to say that as a society we tend toward binaries, the black or white, and do poorly with messy middle spaces. Nature itself is regarded as sublime or corrupting; humans are assessed as either men or women, with the attendant expectations and norms. The connection is not superficial. In *Queer Ecologies*, editors Catriona Mortimer-Sandilands and Bruce Erickson note, "The powerful ways in which understandings of nature inform discourses of sexuality, and also the ways in which understanding of sex inform discourses of nature; they are linked, in fact, through a strongly evolutionary narrative that pits the perverse, the polluted and the degenerate against the fit, the healthy, and the natural."[27]

In other words, when it comes to fungi, how we regard them says something about the role we see for them in the world, a view that has social implications. The field of "queer ecology" in part explores the social assumptions that have been justified on the basis of observations in nature, chiefly as perceived by white men in the sciences, and emerging in parallel with the rise of feminist and queer theory. These assumptions are challenged by such organisms as slime molds and fungi, which serve as binary-bending lenses and avatars in science fiction, such as in the work of Rivers Solomon and Octavia Butler.[28] Feminist theorist Donna Haraway describes the interspecies relations among humans, fungi, and by extension all living things, as the interactions of "queer messmates." "I am not a posthumanist," she writes. "I am who I become with companion species, who and which make a mess out of categories in the making of kin and kind. Queer messmates in mortal play, indeed."[29]

The desire to cram an inscrutable life form into a sharply defined box is on display in the very ways we speak about fungi, slime molds, and other "lowly" organisms. This is evident in the uphill battle faced by Dr. Evelyn Fox Keller in getting the scientific community to accept that *Dictyostelium* represented not a slimy fiefdom ruled by "founder cells," but an emergent, leaderless organization. The binary-eliding nature of these organisms, and the culturally subversive dimensions of their nature, may help explain why a 2018 survey found that around 11 percent of American mycologists identified as LGBTQ, nearly four times the national average.[30] (In that same study, 43 percent identified as female, 56 percent as male, reflecting the gender divide within the sciences, albeit a gap that is gradually closing.[31])

"I'm sure as more and more young people come up in the ranks, that number will continue to grow because right now the mycological society is overwhelmingly old white dudes," Kaishian told me. "I think a lot of mycologists, even if they don't have an identity that is marginalized, they sort of find working with organisms that are really kind of hated and feared and regarded so poorly, you kind of develop a little bit more of this compassion, and a little bit of this sense of, like, 'Hey, like, why are we thinking like this?' You start to challenge the paradigm. It feels like it's a space that has a growing power that's collective, and expansive, and inclusive, and exciting."

Kaishian, who after her paper was published accepted a postdoc position as the curator of fungi at Purdue University, holds out hope that there may yet be a way to "reclaim" the terms by which we understand fungi and their study, not entirely unlike how the word *queer* was gradually reclaimed by the LGBTQ community.[32] To be clear, this is not meant to conflate the experience of the LGBTQ community with field mycologists who, again, tend to be white males for whom we can assume the experience of cultural marginalization is generally unfamiliar (except, perhaps, for the slight awkwardness of being known among friends and colleagues as "the mushroom guy").

Nevertheless, mycology is clearly drawing in people from ever more diverse backgrounds and life experiences; like culture at large, that might present an opportunity to shift the greater scientific endeavor toward something a bit more inclusive of different points of view, an expanded vernacular of ways of being and knowing alike. What if the "queer science" of mycology or the "nonbinary" nature of fungi themselves offers unique opportunities for enfranchisement in the broader scientific project, for communities that once felt excluded or marginalized from those spaces?

Kaishian, for her part, is trying to bring a greater consciousness about inclusion into her work with fungi. Interestingly, her job as a taxonomist is inherently one of drawing distinctions along binary lines. The very tool used for assessing the identity of a mushroom is called a dichotomous key; between two possible traits, you choose one and move to the next pair, gradually fractioning your way to the "correct" identity.

But it doesn't have to be that way. In her own taxonomic work, Kaishian said she hopes that she could perhaps "reclaim" it from the parochial, patriarchal proclivities of "capital *S*" science, and encourage a reconsideration of what such efforts at classification really represent.

"For me, it's not like a possessive thing, but it's giving them recognition," she said. "They've been on this multibillion-year journey to be where they are now, and we're here together in this snapshot of time. And it's like, no matter how small they are, and how obscure or how removed they may be from daily life, they *exist*. To give them a name and to just honor their existence, that's my motivation."

CHAPTER TWELVE

Who Speaks for the Mushroom

Donald Moncayo turned off the highway and took us down the lengthy flank of a Petroamazonas refinery. The withering thunderstorm we'd braved in our drive through Shushufindi had started to dissipate. From the back seat of the pickup truck, I could see a gigantic flame dancing in the air just behind the trees, reflected in a glittering constellation of raindrops on the windshield as the conflagration drew nearer, looming larger. I had heard about the mechero, Spanish for "lighter," and despite having already seen pictures of them, to finally behold one with my own eyes left me awestruck. Margaux, a French volunteer for the Union of Affected People Against Texaco, or UDAPT (a Spanish acronym), and our impromptu translator for the "toxic tour" that Moncayo led on behalf of the organization, turned to address me as I sat in stunned silence. "I didn't want to tease you," she said, "but it is really insane."

We parked about thirty meters away, a distance at which opening the door still invited a roaring, wind-warping wave of heat. Across a dirt lot, a pair of thirty-foot metal pipes protruded straight out of a rectangular pit. Raindrops were still falling as jets of natural gas burst upward into a giant forked tongue of flame that writhed above our heads like an angry chained tiger. It appeared to me as the absolute negation of a tree, two black trunks drawing crowns of flame up from an empty pit in the earth, casting an orange halo upon the dense rim of Amazonian canopy that surrounded it. All around the base of the mechero were dark heaps, the singed remains of thousands of insects—beetles, butterflies, moths, grasshoppers, others I'd never seen before, some as large as birds—every night drawn by the thousands into fatal proximity with the brightest light in the rain forest.

The mechero was just one of 384 spread throughout the Sucumbíos region, a half hour's drive from the border with Colombia.[1] Instead of being sold, the natural gas was simply burned, twenty-four hours a day. Or, if a strong-enough wind blew them out, vented into open air. Some sit mere paces away from schools. Once opened, they never stopped spewing; this one, Moncayo told me, was first lit in 1972. Elsewhere on our "toxic tour," we visited the first pumpjack installed near what was formerly the riverside home of the Indigenous Cofán people. As of the early sixteenth century, as Spanish colonists first arrived in the region, there were as many as thirty thousand Cofán living in the area. By the early twenty-first century, that population was less than 1,500.[2]

In 1967, Texaco struck oil near what is now called Lago Agrio, a name that means "sour lake," named for a town in Texas where the company made its first big break.[3] It's also called Nueva Loja, named after the southern city of Loja, established as part of a government redevelopment effort to make "nonproductive" land more profitable, leading to mass settlement of the area shortly before the discovery of oil.[4] By 1972, Texaco was pumping crude through a vast network of pipelines that perforate the landscape en route to a port on the country's west coast. In the decades since, the long, inevitable series of profligate oil spills have turned many parts of the verdant landscape black and toxic—in waterways, atop and deep beneath the soil—as soaring cancer rates, economic and social disparities, and ruthless legal maneuvers have left the local populations with the mess of extractive industry and little, if any, of the economic benefits. UDAPT insists that Chevron-Texaco disposed of some 650,000 barrels of crude oil, and more than 16 billion gallons of toxic formation water, into the soils and waterways of the Amazonian rain forest. The decades-long injustice has been referred to as "Ecuador's Chernobyl."[5]

On a farmer's property overlooking an Amazonian valley, Moncayo plunged a core sample of soil where another pool had been "remediated" by the state oil company. Following a few cylinders of brown clay came chunks of gray slop that reeked of petroleum hydrocarbons; Moncayo, wearing a shirt that said CHEVRON TOXICO, made a practiced face of disgust for my camera. Just downslope from the contaminated soil sample, a black stain in the mud. Striking it with a machete revealed the hardened oil that broke in flakes like flinty obsidian. If I weren't aware of what I was looking at, it might have even been beautiful.

Later, after a river bottom–scraping ferry ride across the Río Aguarico, we visited a piscina, or pool, that sat just downhill from a defunct oil well. Moncayo waded into the black muck—a combination of petroleum, adhesives, and other industrial offal—on a floating walkway of semisubmerged sticks, reaching down to pull a pole out from the muck that quickly revealed itself to be some twelve feet tall, dripping with black sludge. A drainage pipe led from the piscina to a ledge; when it rained—which in the equatorial Amazon is every day—the pool would inevitably overflow. Downhill, greasy stains confirmed that the pipe didn't discriminate between oil and water. A black handprint was marked on the trunk of a tree near the shore of the piscina; it was all meant to be photographed, shared, seen. Chevron had taken to denying the reality of this situation, along with its responsibility.[6]

Since the departure of Texaco in 1992, this unlined pit and more than eight hundred others like it simply sat where they were, as the oil companies and the Ecuadorian state pointed at one another as the party responsible for cleaning it up. In 2011 the Sucumbíos provisional court ordered Chevron to pay $9 billion in restitution for the thirty thousand plaintiffs from communities impacted by the legacy of extraction, of which these examples represented the barest fraction. The company simply refused, and a tangled, vicious series of international lawsuits, countersuits, arbitrations, and other maneuvers has held things up ever since.[7] Meanwhile, the pools remained; the oil stayed in the soil.

Moncayo estimated that in the decades since the piscinas were first established, thousands of cattle and other animals had fallen into them and died. It was only because I was watching my step to avoid the same fate that I saw the little yellow mushroom growing up from an oil-soaked log.

The bus from Quito to Lago Agrio took about eight hours. Out the window, the Amazon unfolded endlessly in all directions as the bus traced winding, uneven roads alongside the sheer rain forest valleys. A few hours out from Ecuador's capital city, the pipelines started to come into view, growing more regular with each mile, gunmetal stitching in a landscape that was otherwise a thousand shades of green. I wept in the bus.

When I finally stepped off on the outskirts of Lago Agrio, my eyes immediately stung, not from tears but from the air. I'd missed my stop in the center of town, and texting my host, stood waiting on the dusty edge of a busy two-lane highway. It was just a few minutes before Lexie Gropper rolled up in a well-worn compact truck. After I hopped in, we drove through the center of Lago Agrio en route to Amisacho Restauración, right in the heart of Texaco's extractive legacy. The area had been settled since the mid-'60s, originally as a base camp for exploratory drilling operations; a city built on oil, in every sense. Pump 26 was the address Gropper gave to anyone trying to find the fourteen-hectare, formerly deforested cattle ranch–turned–reforestation project. To Gropper, it was also home. She unlocked a gate, and we passed the defunct pump—resembling a red, steel crucifix—as we drove through and down a steep, densely forested ravine. "Welcome to a little paradise, in hell," she said, parking next to another white van with pictures of mushrooms on its side. CULTIVO DE HONGOS COMESTIBLES, it read.

Just shy of thirty, Gropper is fluent in Spanish as well as soil biology and social justice discourse, with a bottomless well of enthusiasm for solving the seemingly intractable challenges facing her adopted home. She moved to Lago Agrio from Atlanta, Georgia, in 2014, as a project coordinator with the Amazon MycoRenewal Project, since renamed CoRenewal. Its work in the region had been focused on cultivating fungi that could break down the oil and help to counteract the contamination left by more than fifty years of oil extraction. From what I'd seen, mushrooms seemed content to grow even in areas deeply soaked in oil; some of the hardiest petroleum-degrading species of fungi in the world might be found in and around Sucumbíos.[8] Those could, at least theoretically, then be deployed at scale to kick-start trophic cycles and a gradual recovery of local ecosystems.

"It never actually went into the field," said Gropper of the project she was previously involved in. "The same species that you grow out onto a plate and keep isolated, they don't have a chance in the Amazon. So I'm interested in other, dirty ways to expand species, and create environments where the natural diversity of organisms are going to proliferate regardless."

Gropper told me this as we stood outside of mushroom lab that she maintained with her husband Luis Muñoz, a cheery mechanical engineer turned reluctant mycologist, as Gropper described him, whose family owned Amisacho. Inside the lab were racks of mushroom bags and inoculation materials for the medicinal reishi and turkey tail mushrooms they still cultivated. Outside, gleaming pressure vessels sat under corrugated plastic awnings, mere feet from misty, thickly interwoven rain forest canopies where monkeys swung from tree to tree. Maintaining a variety of fungal cultures for food production in this hyperdiverse, humid environment had proved too much trouble.

"As a mushroom cultivator in the Amazon—and we're the only mushroom cultivators in the Amazon at this elevation that I'm aware of thus far—it's been extremely challenging, because everything is so diverse. And so finding resilient species that aren't dominated and can colonize the substrate quicker than something else is difficult. It's kind of reiterating that idea of monoculture."

The themes of monoculture also resonated with Maya Elson, formerly of the Olympia Mycelial Network and Radical Mycology Collective, subsequently the executive director of CoRenewal. She also recognized the limits of the "just add mushrooms" approach in addressing problems of the scale and complexity represented in Ecuador. In fact, in the very language of the common mycoremediation approaches, she saw an echo of the extractive colonialism that had caused the problem in the first place.

"It's like, 'Let me just kill all of the native microbes, and then colonize them with my white oyster savior mushroom,'" she told me. The mycoremediation nonprofit remains active in the Amazon, investigating native microbial communities in hopes of informing future remediation strategies, looking for native fungi and other microbes that degrade the petroleum. Working with UDAPT, they plan to establish sample plots in oil-contaminated sites, identify the native petrophilic (oil-loving) fungi, such as *Pleurotus*, in order to culture and propagate those with the most effective enzymes for breaking down oil.[9] "Oyster mushrooms have incredible abilities to biodegrade petroleum, but I strongly believe we need multi-trophic approaches to bioremediation, and we need multi-successional approaches," said Elson. "Ecological *and* social diversity inclusion are key. It's not necessarily helpful to bring in a non-native mushroom, call it a day, and say, 'All right, we fixed the oil spill!'"

Frustrated by the slow pace of progress in earlier mycoremediation efforts, Gropper gradually found her focus shifting from oil to soil, and its dynamic relationship with the people who lived upon it. The cultures that might stand the best chance of restoring the soil, she had begun to realize, might not be fungal, but human. Long before the days of Texaco, human beings left deep, black deposits in the Amazonian earth. But it wasn't petroleum. In fact, it was some of the most fertile soil in the world.[10]

Terra Preta (Portugese for "black earth") was discovered in the Amazon in the 1960s, marking a dramatic shift in the modern understanding of the agricultural history and potential of the region. For many years, it had been taken as read that the region was simply too hot, humid, and the soils too poor for the establishment of large, complex civilizations.[11] It was unclear how deliberate the deep, dark bands of carbon rich "biochar" actually were, filled with remnants of charcoal, bone, pottery shards, chicken droppings, and other signs of human oversight. Carbon dating and archaeological evidence suggest the processes that created the deposits first began as long as seven thousand years ago.[12]

The biochar of Terra Preta is, first and foremost, a rich sink of organic carbon. Organic carbon is any carbon-based molecule that originates from something once living. It leaves a black or dark brown color; typically, the "oxisol" soils of the Amazon are red, yellow, even white, the heat and humidity making it so that organic carbon decomposes too quickly to accumulate in soils, converting instead into carbon dioxide, leaving behind mineral sand, silt, and clay.[13] By contrast, in addition to carbon, biochar soils show enhanced nitrogen, phosphorus, potassium, and calcium content.[14]

There are techniques for encouraging carbon richness and soil structure, such as in Korean Natural Farming.[15] Some call the approach used in forming Terra Preta traditional ecological knowledge or TEK.[16] At the base of these processes is the burning of organic matter, predominantly wood, in a manner that doesn't burst into open flame, but instead smolders in a low-oxygen environment. The resulting char has a microscopic honeycomb structure of organic carbon—a veritable sponge for water and nutrients, and an ideal habitat for microorganisms with a high cation exchange capacity, to boot; that is, the ability to hold ions such as ammonia and ammonium, and pass them on to plants and microbes. One need only activate the biochar with nutrients (essentially fermenting it); compost and urine are commonly used for this purpose.

Uphill from the Amisacho lab, the sound of angle grinders echoed through the trees. Muñoz and Amy, a visiting childhood friend of Gropper's, were hard at work, casting off arcing sprays of orange sparks as they cut vents into the tops of a pair of fifty-gallon barrels. These were their scavenged biochar furnaces, low-tech generators for high-TEK soil. Around the corner in an open dining area rented out for large gatherings, Gropper kept jars of soil used in "shake tests" to visually assess their structure and the presence (or lack) of organic activity. The variations were stark: thin, light brown and striated; thick with clay; or yellow and hazy. Each represented a member of the community who was just starting to assess their soil's health, the first step toward improving it.

The soil samples were part of a permaculture-based bioremediation curriculum that Gropper was helping to develop, in partnership with UDAPT and a local NGO called La Clínica Ambiental. The latter organization's mission, as Gropper describes it, is to advance "social approaches to ecological reparations," mainly through free permaculture courses offered to affected farmers and communities. Early in 2019, they held the first Environmental Reparations Assembly, a gathering of local leaders and international experts in bioremediation, that together decided the best way to move the ball forward for local remediation efforts was to first educate affected communities in the practices of permaculture. Designed with the help of ecological systems designer Nance Klehm, the Guardians of the Soil course that emerged aimed to put into the hands of residents living on or near contaminated sites tools to assess the quality of their soil without the aid of difficult techniques or sophisticated equipment that wasn't forthcoming anyway.

"I'm fascinated by industrial bioremediation, but is that replicable and applicable here?" posed Gropper. "No. You need a lot of money in order to be able to contract somebody to remediate everything that's going on here. And this is the poorest region in Ecuador. So, you know, we could bring in bioremediation and patents and specialists like that and do a pilot project and demonstrate that it works, and then that's going to sit there forever, because there's not going to be the financial resources to repeat it."

Gropper was born and raised in Atlanta, Georgia, eventually moving to the Appalachian Mountains of Boone, North Carolina, to study ecosystem sciences. There, she developed her mycological knowledge, before getting a job with the mycomaterials company Ecovative Design near Albany, New York. Afterward, she took a job with CoRenewal, landing her in Lago Agrio where she met and fell in love with Muñoz. Having become personally invested in the region and its people, Gropper told me "I wasn't able to just come and go like another volunteer."

The majority of the thirty thousand claimants in the UDAPT case against Chevron-Texaco were small farmers and Indigenous communities.[17] The contaminated environment had been a fact of life for as long as many of them had been alive; almost none of them had the resources to conduct an analysis of their soils, a situation not helped by the rampant use of pesticides and chemical fertilizers.

The hope was to provide tools for local farmers to assess the state of their lands, which would inform subsequent stages of remediation and a broader move toward regenerative agricultural techniques. The bag of sensory tricks available for assessing soil health was described by one of the teachers involved in the project, Ignacio Simon, as the "microscopia de campesina" (the farmer's microscope). "It's in using all these different methods, and at the end of it, you're going to have a really good idea of the reality of your soil," Gropper said. "How healthy are your plants? Are they photosynthesizing? Are they yellow? Is the soil compact? It's just simple landscape reading, which is complex."

Other projects reinforced these efforts. For example, the new network of permaculture farmers that the coalition was helping to foster were the backbone of the Traveling Hope Fair. The fair traveled to local villages and shared some of the only organic produce and products in the province, raising awareness about food sovereignty and holding workshops to demonstrate the various income streams that were possible with their agroecological approaches. A "passport system" had also been devised to facilitate knowledge exchange, providing a stamp for each stop along a circuit of local farmers utilizing permaculture principles.

Instead of working to kick-start trophic cycles and succession in soils directly, Gropper, Muñoz, and their collaborators were attempting to instigate community engagement with the landscape in ways that could accomplish the same overall goal. Beyond damaging the natural nutritive cycles that underpin healthy soils, contamination had also severed relationships between people and their land. Fungi were part of that, but really the work was focused on ecosystem health in the broadest sense, a picture that included fungi among all the other interacting agents in a healthy "soil food web," as soil biologist Dr. Elaine Ingham describes it.[18]

"You need to make sure you have a strong-enough soil ecology present," Gropper explained. "Little by little, building up your soil structure and soil quality, and then working into more perennial plus annual plants, and then involving more mycorrhizal fungi, working with your plants that are getting established. Right now we're just working on gaining confidence and reconnecting with soil. For so many people who've grown up in a contaminated area, they no longer have this connection."

The whining of the angle grinders stopped for the moment, and the din of the jungle returned. Neither the trees nor the spaces between them were quiet or still, resonating with the calls of birds, crickets and frogs, and intermittent sheets of pouring rain. All around was a literal churning life, and Gropper's work seemed an effort to lower the proverbial oar into that flow. It made the notion of "fixing" the environment seem petty next to the goal of getting into right relationship with the cycles of life that carry on all around.

"It's so easy to talk about solutions, but if it doesn't respect the social, cultural, historical, climatic realities, it's not a solution," said Gropper, indicating her frustration with a "solutionist" mindset that offered immense promise but had yielded so little in the way of lasting results, and go figure. "It's a very Western way of thinking."

Amisacho had not given up on fungi, though. Far from it. The reishi and turkey tail they grew, for example, were used to make tinctures, sold through UDAPT to local cancer patients. UDAPT kept records of the cancer patients among its thirty thousand claimants, whose conditions were believed to have resulted from the contaminated soil, water, and air. People living near the oil pits had been shown to suffer from a suite of health problems, in particular cancers. Elevated rates of cancer of the stomach, rectum, skin, soft tissue, kidney, cervix, and lymph nodes;[19] certain cancers were particularly high among children under ten years of age.[20]

The tinctures, which marked Gropper and Muñoz's sole source of income, were reportedly very popular. "The testimonies of the people using them are really profound," said Gropper. "Selling edible mushrooms is really cool, but it doesn't make me feel like I'm serving. So the production of medicinal mushrooms, for one, they're a lot more independent; the harvest isn't just from one day to the next and then they rot. And I don't have to go searching for where to sell a pound of mushrooms. And also I feel like we're serving because it's actually helping people."

The nearest hospital that could take care of cancer patients is the Sociedad de Lucha Contra el Cancer, in Quito, some eight hours away from Amisacho. For many in the area, traveling there is not an option, so the tinctures represent a rare bit of help, and hope. In this context, terms like *medicinal sovereignty* move from platitude to deeply meaningful. Indeed, it was the cancer-fighting potential of medicinal mushrooms that originally got Gropper interested in mushrooms.[21] In this situation, they weren't just a form of "alternative medicine"; for some, they were the *only* medicine, which is why Amisacho does not sell its tinctures outside its community. Also, Gropper says, because "enough has been extracted from Ecuador."

The first time I'd reached out to Gropper, it was to find out about the "oil-eating mushrooms" that had grabbed headlines. Admittedly, I was a bit deflated upon hearing that such projects had wound down at Amisacho. But upon hearing the scope of the work they were undertaking,

I saw my first example of remediation that took place outside of the contaminated soil or water itself. The environment was contaminated as a whole—you can't walk anywhere without encountering it with one of a variety of senses—and of course that environment included the people living there. That contamination was a consequence of human activity, with consequences for human communities, so it seemed right that the remediation should happen in the human dimension every bit as much as in the soil itself.

"The focus is really ecosystem restoration, and recognizing that we also are a part of the ecosystem, no different," said Gropper. "We focus on restoring ecosystems, restoring health, and restoring community, so it's becoming more and more of a complex social cohesion, and that's what really inspires me: diversity in the soils, diversity of humans, diversity of hands and professions, diversity of organizations involved . . . and diversity of gut flora!"

Intersectional Mycology

The sky was overcast and the wind cool as Robert Wallace, member of the Barona Band of Mission Indians, spoke to a group of about thirty people gathered in a courtyard at the EthnoBotany Children's Peace Garden in San Diego's Balboa Park. Every few minutes, he paused to wait out the roar of a jet overhead.

"What I see you guys doing today is really beautiful," he said. "That's why I had to come down here. I don't show up to things that I don't believe in." Wallace related a couple of stories from what sounded like an incredibly eventful life ("I'm just a guy who walks under the right stars at the right time," he insisted), passing pinches of tobacco around the circle before walking its perimeter and shaking everybody's hand, one by one. The first POC Fungi Community Gathering on Kumeyaay Territory was under way.

Across from the garden entrance was a giant mural of King Tut-ankhamun's golden mask, painted on the curved surface of the WorldBeat Cultural Center. Once a million-gallon water tower, it had for twenty-three years served as a hub of intercultural exchange in San Diego, under the leadership of one Makeda Cheatom.[22] Covering the building's curved surfaces were vivid paintings of Egyptian, African American, and various Indigenous cultural figures and symbols. Inside, flags from all over the world were draped from the ceiling, beside depictions of ankhs and pyramids under construction. It was the ideal venue for an event meant to bring BIPOC and LGBTQ folks together, officially around common affinity for fungi. A big, hand-painted banner—or pride flag, as it was described—hung next to the entrance, decorated with images of turkey tail mushrooms and mycelia and the words POC FUNGI COMMUNITY across the top in big letters, complete with a hashtag for social media–ready selfies. Running down its sides were symbols and figures representing BIPOC culture and contributions to the modern understanding of fungi: Maria Sabina, George Washington Carver (widely underappreciated as a mycologist), a Kaminaljuyu mushroom stone, Glass Gem corn sprouting huitlacoche fungus.[23]

Mushrooms and fungi were the subject at hand, but the true focus was much broader, and deeper. More than a place to talk about mushrooms, it was an all-too-rare opportunity to gather for a community whose members reported feeling unwelcome or marginalized in many mainstream spaces.

"We all know that this isn't just a mushroom event," said the lead organizer, Mario Ceballos, taking the stage to kick off the day's series of lectures and workshops, conversations and performances that would extend into the evening. It was clear from the outset that he was speaking not to a room of strangers, but to his community. "As much as we are fungi-loving folks, we come with a lot, right? And what's good about this space, we don't have to leave none of that at the door. We come in here, we can bring it, we can bring our narratives, our story, our experiences, whether they be good or bad."

By midday the huge, colorful chamber was completely packed, the grounding aroma of sage suffusing the air. Many visitors sat among the rows of folding chairs, listening as each speaker and panel took the stage; others milled about the various tables and the small café. A majority hailed from nearby areas, as revealed by a show of hands.

Ceballos, a robust, jovial Chicano wearing flannels, a ponytail, and glasses, recited the mission statement of the event that he had nurtured into being from an idea raised about a year prior in conversation with members of his community. "The POC Fungi community is a safe space," Ceballos read, "for queer, trans, Black, Indigenous, and people of color to learn, build community, and reconnect with nature, and do it all in a space where we are free from being objectified, tokenized, silenced, or any form of oppression."

Everybody passing the registration table had been required to sign a code of conduct forbidding oppressive, racist, or otherwise other-izing behavior, which along with a sliding scale admission fee and pointed land acknowledgments seemed to have become a feature of these emerging social justice and inclusion-minded fungi gatherings. The logo for the POC Fungi Community event, also depicted on a widely distributed zine of the same name, was the silhouette of a mushroom made entirely from a stack of words, including: SOCIAL JUSTICE, REPRESENTATION, CULTURE-FOOD SOVEREIGNTY, INTERSECTIONALITY, RECLAIMING.

Visitors perused tabletops stacked with various craft items, along with many familiar fungal medicinal products: tinctures and stickers and shirts and posters and bags of dried, red-brown slices of reishi. A copy of the Mayan Codex was laid out on a table, while in the corner, a tattoo artist was on hand for anyone in the mood for a subdermal souvenir, busily buzzing throughout the day. Between the scheduled events, folks conversed, perused the tables, and visited with the five puppies that someone had brought to the kids' corner, to the particular delight of the adults, maybe most of all Olga Tzogas. "Why are you *perfect?*" she gushed. "This is the new standard. If there's no puppy, it's not a real fungi gathering."

Tzogas had driven all the way to Southern California from upstate New York on a weeks-long road trip in her trusty van, stuffed to the gills with the Smugtown booth and its many accoutrements. Among the familiar faces was William Padilla-Brown, flashing new gold grills and a bright white hoodie splashed with his company's neon-orange *Cordyceps* logo. Tzogas presented a talk about the mushrooms of her ancestral Greece; Padilla-Brown gave a talk about citizen science (his portable gene sequencing rig was on display), before switching gears from soft-spoken science nerd to psychedelic hip-hop emcee, closing out the event with an energetic set.

At the heart of it all, though, was the theme of medicine, a touchstone in literal terms, and in a broader sense of healing. Psilocybin was at the center of discussion, along with the attendant issues of extraction and access that often come up in relation to communities of color, in particular to Indigenous folks with ancestral relationships to plant medicines. "When we talk about these things, we often talk about it in secret, in soft voices, because it's that sacred to us," Ceballos told me. "So yeah, it bothers us to see our medicines being commodified. Yeah, it bothers us when we see scientists pulling up our medicine and reducing it down to compounds, squiggly lines, telling us maybe that medicine isn't even *for* us. Because when the retreat centers open up, I don't know if I'd be able to afford five hundred dollars a session."

At the age of sixteen, Ceballos had a life-altering experience with psilocybin mushrooms that, he recounted, brought him to a place of forgiveness and understanding with regard to his family and their fraught history. It began a process of reconnection with forms of medicine to which he felt deep cultural resonance, but also social disconnection; largely a result of its criminalization. After that formative experience, his mycological interest began to soar at the same time as the popularity of psychedelics therapy also began gaining steam in the mainstream. This confluence is what provided the impetus to form the POC Fungi Community. "Decriminalization was coming around the corner, and I seen what was going to happen, I already knew it," he told me. "I needed to be preemptive."

In consultation with community members and elders, Ceballos had managed to establish a rare space where the perspectives of LGBTQ and BIPOC folks would be represented in the conversation about plant medicines, free to openly and honestly address the mounting matters of access and exclusion, criminalization and commodification. On this first attempt at coming together, they had evidently tapped into a massive well of interest.

Word of the convening had even traveled outside the country. An announcement was made that the Mazatec community in southern Oaxaca had reached out to extend their encouragement and support for the gathering. Paula Graciela Kahn, who had led the "Justice for Psilocybina" panel at the Radical Mycology Convergence in Oregon over a year prior, recited the statement from Inti García Flores, a researcher of Mazatec culture in Oaxaca, at the beginning of her presentation. "Hopefully, the purpose of healing humanity will be fulfilled," it concluded, "so our world can change and transform in consciousness."

As the profile of mushrooms grew, the hope at the POC Fungi Gathering was to steer and steward the emerging discourse in the right direction. It was also to establish a seat at the table for communities that are historically likely to be excluded. Indeed, mushrooms and other traditional plant medicines are criminalized in ways that disproportionately affect marginalized communities. In some cases, this amounts to simultaneous criminalization and commodification; there are people in jail for partaking in practices that can be traced back to their direct ancestors, while a person without those cultural connections can avoid such consequences because they have approached the same medicine with profits in mind. As journalist Kaitlin Sullivan put it, "When people of color make their illegal drug use known, rather than being rewarded with a book deal, they're statistically more likely than white people to face jail time."[24]

In recent years, figures such as Kilindi Iyi and Kai Wingo—also on the flag—made strides in representing their communities in the discourse about psychedelics, and their historical meaning to communities of color.[25] Tragically, Wingo and Iyi passed away in 2016 and 2020, respectively, but others are continuing to create spaces for BIPOC communities in the discourse about fungi and psychedelics, such as Oakland Hyphae and the Detroit Psychedelic Conference. And as we have seen, fungi offer opportunities to independently grow food and medicine, heal landscapes and waterways, kick-start local economies, and serve as living examples of equitable, reciprocal relationships with nature; conversations about these issues, though, are still very much shaped by the social status quo.

"Nobody's really created a space for us," Ceballos told me. "I've gone into mushroom spaces, and try to bring my narrative and my policy to those places, and they don't want to hear it, you know? So, after that, why am I fighting to get into these spaces, into these institutions and into these conferences, if they were never were built for us in the first place?

They didn't have us in mind. We weren't a priority, you know? POCs, Indigenous, queer, undocumented peoples' stories and narratives weren't ever in the center. So, why don't we create our own space?"

By the afternoon, that space was filled to capacity. Doğa Tekin led a workshop inspired by Robin Wall Kimmerer and the language of animacy, explored through the act of drawing mushrooms. Esteban Orozco, a holistic nutrition coach, spoke to the issue of decolonizing foodways; he and Ceballos had met at the New Moon Mycology Summit (further proof of the principle of "spreading spores"). Other events on the day's schedule included "Decolonizing Turtle Island Food Ways," "Decriminalizing Psychedelic Mushrooms," "POC Inclusion in Peer Integration." Booths were set up by Brujitx del Barrio, Queering the Path, and Decolonize Mycology. Outside, people formed a line for vegan mushroom pozole provided by Veggie Mamacitx.

Sneha Ganguly, who had flown in from New Jersey along with the colorful pride flag that she made, also oversaw a table where she was selling homemade mushroom tinctures. The bottles were labeled with homemade paper that the industrious Ganguly had manufactured herself, from fungal fibers and dyes derived from cinnabar polypore, an example of the potential of fungi in reforming fibersheds, the often wasteful and extractive systems and practices behind the production of textiles. Nearly everything on the various tables had been produced from within and for the benefit of the community; in this way, they were said to have been decolonized.

I bought a turkey tail tincture from Tzogas and a reishi bag from Ceballos to send to my ailing mother. In the final stage of her own struggle with cancer, she had been consuming some anonymous mushroom supplements bought off the TV, which struck me as playing into the growing trend of pharmaceuticalized fungi, sold to a largely myco-illiterate public. Even without any traditional connection to plant medicines myself, by this point I had become convinced that a shard of homegrown, unadulterated reishi tossed into a glass of tea offered more benefit than a processed, anonymous powdered supplement that CONTAINS BASIDIOMYCETE FUNGUS, as the bottles indicated. Despite my own sense of alienation from the medicinal traditions represented at the gathering, I found myself placing real value in the spirit with which it was produced, and got a virtuous feeling from acquiring homegrown medicine for the benefit of a close—indeed, my closest—loved one, produced and acquired with intention, and sold to support a community rather than a company or an industry.

The many windows that fungi offered onto issues of importance to that community quickly became clear. Jiapse Gomez, a young regenerative gardener, took the stage to offer a survey of the basics of fungal biology and mycoremediation for the benefit of the many people in the audience who weren't yet familiar. The presentation was woven through with a sense of social consciousness, aided by Gomez's flair for wordplay, deploying phrases like "It's an ecosystem, not an ego-system!" and referring to fungi as the "great decayers," a nod to a Denzel Washington film, *The Great Debaters*. He made the case for mycoremediation in contaminated gardens, a problem this community faced more acutely than many.

"I think that mycoremediation and bioremediation in general is a very beautiful technique for us to be able to connect with the land again," Gomez said. "This is ancestral knowledge, working with the plants, working with the fungi. These are things that people have been doing for several centuries, so it provides an opportunity for us to reconnect to that type of medicine. But it also provides the opportunity for us to physically clean our neighborhoods that are suffering from these environmental contaminants."

Valerie Nguyen, one half of Fungi for the People, briefly introduced themselves. "My pronoun is *we*," Nguyen said, suggesting, perhaps intentionally, what a *fungal* pronoun might also sound like. It was also a reference to Ja Schindler, representing the other half of Fungi for the People, who could not make the trip from Oregon, where the pair honed and taught cultivation techniques and organized workshops and education programs to broaden access to the skills needed to grow and sell mushrooms. Outside, Nguyen had set up at a table where people were invited to "demystify" the Latin names of mushrooms—often an intimidating obstacle to engagement for the myco-curious, not to mention being freighted with the colonial connotations of Latin binomials—by helping visitors decide on a mushroom name for themselves.

At a breakout group following another talk, Allison Shiozaki referred to an experience at a mycology festival in the Bay Area, where she had interactions that were described as highly toxic. "With mushrooms, or with people?" came the natural clarifying question, along with a laugh, because everybody knew that she meant the latter. "With people." Shiozaki had helped to organize a new group in Berkeley, called Decolonize Mycology. The earlier interactions had been intense enough to settle the decision to remain exclusively a group for BIPOC and LGBTQ mycophiles. The group had announced that, if they were to again hold events that included cis white men, it would be in the special and explicit context of allyship. Hearing these things, it began to sink in just how much I took my access to, and comfort in, these spaces for granted.

The only occasions in which I have personally encountered toxicity around mycology occurred in social media forums, where a distinct strain of chauvinism can seem to run more or less rampant. For instance, one will often see what have come to be called "kill shots," massive piles of mushrooms stacked up like the spoils of a hunting expedition. Those concerned with conservation and a reciprocal relationship to fungi and ecosystems—often women, I've noticed—may comment that these images are extractive, wasteful, or disrespectful of nature. These critiques are often met with rather mean and sometimes vile remarks—often from men, I've noticed—and even suggestions of violence, along with what may be described as the weaponization of scientific language, by way of claims that such concerns about sustainability and respect for the organism come from a place of ignorance about "how fungi actually work," a charge I have seen leveled at PhDs in mycology. And that was just online. For many at the WorldBeat Cultural Center that day, the codes of conduct we signed upon walking in weren't just a performative gesture; they were informed by lived experience.

Somehow, advocating for fungi and their right to exist resonates with a related struggle faced by marginalized people. Little wonder mushrooms could be central to a community's effort to claim space for itself.

"Fungi is intersectional!" Ceballos reminded me. "A lot of people don't want to hear it, 'cause they're, like, 'dude, I'm here for the mushrooms, I'm not here for the politics, colonization, and ancestral trauma,' you know, that seems like way too much work for most. And, well, for most of us who see it, we see them all together. We're like, no, it's all on the same hand, actually, not even a different hand: mushrooms have been commodified, demonized."

The gathering in San Diego—that is, unceded Kumeyaay Territory—struck me as a compelling example of what fungal fellowship could look like. More than a meeting of mushroom nerds, it represented a moment of intercultural solidarity, as well as, perhaps, a kind of inter*species* ally-ship.[26] The fungi had brought together folks who placed the land front and center, as well as their own communities' struggles, aspirations, and traditions, and in so doing seemed to have facilitated an opportunity to do some much needed healing.

"Hey, we're remediating, right?" Ceballos had chuckled as he neared the end of his speech. Despite the sense of humor, I didn't get the feeling that he was joking.

The POC Fungi gathering was the last event I physically attended while writing this book. About two weeks later, a deadly pandemic swept through the country and the planet, calling into question the meaning and relevance of what I'd been documenting over the previous months. Of what value is citizen science in a time of mass evictions and unemployment? What good is a fungi festival, however socially engaged it may be, when people can no longer safely gather in groups? Mushrooms might represent promising forms of medicine, sure, but is that really worth much in the context of a worldwide pandemic? For Ceballos and his community, the value of the fungi wasn't questioned; they had already been helpful in fostering a greater overall resilience in his community.

"After the shit hit the fan, I've just been proud of the community, proud of the response, proud of how we were able to mobilize and orga-nize a lot of mutual aid," Ceballos told me when we spoke some months after the lockdowns began. "The gathering really did provide a safety net, a sense of security, a sense of, like, the community's got me. And not just the community, but maybe these mushrooms got us, too, you know?"

Before COVID, the community in Kumeyaay held an open-air mar-ket on Sundays called Tianguis de la Raza, right next to the WorldBeat

Cultural Center where the fungi gathering took place. At the beginning of the outbreak, the tianguis was put on hold over safety concerns. But after a months-long hiatus, and much debate within the community, the decision was made to open it back up, taking appropriate measures to remain safe, of course.

"There was a consensus reached on autonomy; if we are relying on the government, if we are relying on these systems of oppression to let us know when it's okay to gather, when it's okay to move about, are we truly dismantling these systems and truly creating autonomy?" said Ceballos. "That was one of the main things that came out of the gathering: Why don't we put faith in ourselves? Faith in our medicine? If we're truly building autonomy, we come out and we do it in a safe way, keeping public health in mind, of course, always. All the Indigenous communities, we've felt and seen this before, especially here in Kumeyaay Territory—they survived four genocides before—so when Indigenous tribes with this pandemic, you know, we've been victims of this before, so we're not taking it lightly—we're taking it very seriously—but at the same time, we know best what's good for us."

The same question may come up for you that has come up for me throughout the process of reporting and writing this book: What does this all have to do with fungi? It's easy to stay focused on the mushrooms, fascinating as they are, but perhaps it's what they facilitate that is what makes them so powerful, and so prone to bring people together around them, from so many different walks of life, for so many different reasons.

At this point in the book it would be natural to invite you to cast your mind to the future, to imagine what fungi might help us accomplish that we otherwise couldn't. Mushrooms are going to save the world, we're told, ushering us into some kind of mycotopia, rich with ecology-restoring mycotechnologies and a resilient, equitable mycoculture. And as we've seen, there is much that fungi can do for us: providing food and medicine, motivating local economies and conversations about social justice, and helping to repair damaged ecologies. However, I am convinced that the most promising power of fungi lies in their apparent power to bring people together, and to shift perspectives.

Yes, fungi offer many exciting "applications." It's thrilling to imagine mycelium-grown houses; sustainable burial practices using mycelium–infused "death suits"; small mushroom farms distributed throughout neighborhoods to provide food and medicine while digesting waste; psilocybin-derived therapies that transform lives and comfort the terminally ill; mycoremediation and -restoration approaches that may help us rebuild soils or foster sustainable local economies. Companies like Ecovative and MycoWorks and Emergi produce incredible mycomaterials that seem poised to replace styrofoam, leather, and meat, respectively. Such innovations and the visions of the future they inspire are compelling, encouraging, and inch closer to reality by the day. They may even be necessary, but I doubt they're sufficient.

Although the technologies involved have improved, and the need is more dire than ever, these sorts of solutions, when you come right down to it, aren't really anything new. Over fifty years ago, for example, yeasts were entering an era of industrial-scale production; praised as "the next source of protein"; prognosticators of the time imagined vast facilities laden with vats that would produce enough yeast-based nutrition to meet the needs of the entire world, using a fraction of the footprint required for agriculture and livestock.[27] Yeasts are indeed employed at unprecedented scales to the benefit of many, but they certainly did not become a solution for global food shortages, like a sort of Soylent Beige. Incidentally, more recently a meal replacement product called Soylent was hailed as the next era in food after it launched in 2013, but so far it has yet to represent much more than another Silicon Valley hype cycle.[28] The inconvenient truth is that under our current social and economic order, market logic determines and often limits the scale and distribution of any such solutions.

We hear similar promises made about fungi and applied mycology as we have about the internet. The emergence of computer networks did indeed transform the world, but largely in accordance with the very market logic that enabled their proliferation.[29] Social media turned human beings and our relationships into units of engagement to be aggregated and monetized, our personal, social, and even civic life constrained and even warped according to what drives advertising revenues. However much the internet might resemble a mycelial network, the two do not operate the same way. The latter tends to distribute resources in a manner that benefits the ecology in which it's enmeshed, at least in the mycorrhizal examples we would most want to take as models for our own society. It is, generally speaking, a distributive being that functions as part of an interdependent, co-constructed web of life. The internet, by contrast, has come to operate mainly to the benefit of the few who own the network.

Really, the promises of game-changing innovations almost never quite seem to fully materialize to the benefit of the world at large. For all the speed and efficiency that we celebrate in our computer technology—we don't use quite as much paper these days, right?—our deepening dependence on electronics is simultaneously degrading the planet and its ecosystems with what is now the fastest-growing of all waste streams.[30] That's not to question the potential of computers and networks and human-made systems to do good things in the world; the problem is the context, an extractive mindset that shapes and limits the potential of any good ideas. Our landscapes are still dominated by monocrops and factory farms, even as we enter another cycle of hype about "the future of food" coming from a synthetic meat lab, with its own theory about how its product will scale. As new myco- and other solutions vie for public attention and investor dollars, it seems unlikely that they will change anything if simply plugged straight into the same frameworks that have created the problems we face and seek to solve, frameworks that, incidentally, stand athwart the very ways of life that fungi embody and represent. Personally, I'm far more compelled (and convinced) by the notion of a revolution germinating in the community garden or maybe a small mushroom farm in someone's garage than from a start-up's rented high-rise office.

Fungi will undoubtedly play countless crucial roles in our future, if only because they already do, and have done so throughout the history of life on this planet. That may sound flippant. No doubt, there are endless opportunities for fungi to help us accomplish important, practical goals: cleansing waterways; building and remediating soils; degrading our ever-mounting mountains of waste; providing natural, protein-rich food and potent natural medicines; spurring new opportunities to generate value as well as new ways of thinking.

I was—and am!—compelled by the promise of fungi articulated and explored by the people mentioned in this book, among many others. But what use are mycoremediation techniques if those advancing them and advocating for their use have no power to implement them at meaningful scales? What good are meat substitutes at a time when the ecological cliff's edge approaches at a rate that outstrips any hope of ramping up production of these alternatives and mobilizing consumers to adopt them in time? The responsibility is on us, not the fungus, to establish the reciprocal, regenerative future that I assume we all want. As tempting as it is to indulge visions of a fungal version of the Jetsons (I admit, their houses *do* look a bit like mushrooms), it's not very likely. The scale and speed of the ecological and social degradation we collectively face simply will not be met by yet another innovation or technological patch, this time in an organismic shell, to be plucked from the store shelf or ordered from Amazon.

That's a tough reality to accept when the challenges are so dire and so pressing, and our habit of seeking a "fix" so deeply ingrained. Fungi themselves, though, are patient, adaptable, and abiding; they don't repair, per se—there is no 'original state' of nature to which they or we can return—but rather facilitate and actualize. Perhaps what fungal fellowship teaches us is to question our unsustainable systems and change them, not to bury them under mushrooms and hope they'll emerge fixed. It's my opinion that what makes fungi so compelling is not the ways in which we can seize them to our benefit but the ways they can serve as partners and even teachers in the project of realizing more integrative, reciprocal ways of thinking and, therefore, being.

"Despite the undeniable role played by inter- and intra-species competition for evolutionary advantage," writes social ecologist Dan Chodorkoff, "Ecosystem dynamics are best characterized as rooted in the principle of mutualism. Each species plays a critical role in the health and development of the other."[31] That is a description of the world I want to live in; it is also, I think, an inherently anti-capitalist sentiment, and it exemplifies to me what mushrooms demonstrate in landscapes, and in the emerging culture they inspire.

Ultimately the greatest opportunity afforded by fungal fellowship seems one of decentering ourselves and returning to right relationship with nature, which will necessarily bring us into better accord with one another. Such a project neither starts nor ends with fungi, of course, but they offer potent inspiration and, as we've seen, meaningful partnership. Millennia of entrenched traditions govern much of our relationships to plants and animals, and to the more recently discovered world of bacteria and viruses, often approached from an antagonistic or extractive posture, and so it is with fungi; it just so happens that they've been overlooked and reviled for so long that we're confronted with a *choice* about the lessons we draw, and how to regard them as we come to more fully realize their value and importance. Meanwhile, the market is recognizing their value, but my hope is that it does not co-opt or dim the apparent dawning of new ways of thinking, relating, and being that fungi are helping to foster.

In a time of social and ecological collapse—or, to put it more optimistically, *transformation*—we'd do well to recognize and be humble in the face of the reality that, although our bodies take a different form, we are no less embedded within the world than fungi. To act with the health and agency of our environment in mind is to act according to our own best interests, and vice versa. Fungi, in their way, know this, and it's why they've been around as long as they have. If we can accept a similar role in the world, every innovation—technological, social, personal—we will have a proper substrate in which to grow and flourish. Mycotopia, to me, simply means utopia, and it is already all around us. All we have to do is embrace it.

AcknowledgMenTs

When I first stepped into Smugtown Mushrooms in Rochester in the fall of 2015, I had no idea how much it would reorient my life. Rare are the moments that seem to shimmer with such a sense of novelty and possibility as what I felt wandering the forests and poking around the mushroom bags over those two days. A big thanks therefore goes to Olga Tzogas, as the person who first ushered me into this weird and wonderful world of mushrooms and mycophiles. Mush love to you!

Thanks must go to Nick Kaye as well, who approached me in the summer of 2018 with the invitation to propose a book about the strange fungal subculture that I'd begun writing about. He sensed the potential for a book, and I heartily agreed. A massive thanks also goes to Michael Metivier, for shepherding this project from conception to completion, and for showing endless patience and grace whenever I floundered or doubted along the way. Thanks also to Margo Baldwin and the rest of the Chelsea Green team for their support and faith in my efforts. Among the many feelings I've experienced in undertaking this, my first book, is gratitude for being trusted to write it. I sincerely hope I've delivered on that trust.

Thanks go to Ben Goldfarb, not just for inspiring me with his fabulous book, *Eager*, also published by Chelsea Green, its subject being another oft-maligned group of organisms—namely beavers. He was the first person I spoke to as I began to formulate how to approach my own subject, and he gave me crucial assistance in recognizing that, yes, it is possible to write a book.

Alanna Burns has been my mycological fellow traveler from the beginning; we caught the proverbial spore at the very same moment, and I'm grateful for her insight and perspective at various points in the writing process. Josh Viertel will always have my gratitude for introducing me to the editors of *The New Food Economy*, now *The Counter*, for whom my first article was a profile of Smugtown Mushrooms, and where subsequent writing helped lead to this book. On that note, thanks go to Kate Cox, Joe Fassler, and Jesse Hirsch at *The Counter*, for helping me improve and build out my writing on the food, farming, and mushroom beat, and for sending me on some of my favorite reporting adventures.

There is no way this book could have happened without the perspectives, indulgence, patience, feedback, and hospitality offered by folks inside and, regrettably, also missing from the preceding pages. Thanks go to Bryn Dentinger for giving me two full days of his time and insight, introducing me to the work of a professional mycologist, and for graciously taking me on a tour of the mountains, canyons, and diners around Salt Lake City. Giuliana Furci was a source of great insight and encouragement throughout the process of reporting and writing this book. Abrazo! Also thanks to Rikki Longino for hosting me in Utah, taking me on a sublime late-night visit to the shores of the Great Salt Lake, and for introducing me to the local goats and dumpster diving bounty. Lexie Gropper and Luis Muñoz hosted me in Ecuador, one of the most challenging and transformative trips I've ever taken, and I am grateful for their hospitality and humanity throughout. Willoughby Arevalo, Elizabeth Kirouac, and Uma put me up (and put up with me) for three days in Vancouver, which turned out to be a visit that was magical well beyond the mycological. Joanna Steinhardt was generous with her time and insight early on in the process of writing this book, and her own work presented a helpful framework for understanding the DIY mycology community. Craig Trester introduced me to the world of citizen science and ecology in my own environment of Brooklyn, and gave my brain a good stretch each time we spoke. Leif Olsen provided me a cozy place to stay in Asheville, along with always-entertaining conversations, usually over very good beer, that inevitably deepened my mycological perspective. William Padilla-Brown and Cassandra Posey graciously hosted and fed me in their home as I sought to understand their innovative, multifaceted vision of community-level regenerative economies. Peter McCoy was generous with his time and insight as I

ACKNOWLEDGMENTS

sought to understand and articulate the mycoculture he's done so much to build. Sandor Katz allowed me to haunt his home for a couple days as he shared from his endless well of knowledge and insight into the many ways of the microbes, and introduced me to a wonderful community of fermentationists, a life experience I won't forget. Mario Ceballos was a wellspring of support and encouragement as I strove to understand and represent his inspiring community-building work in Kumeyaay Territory. Brian Douglas and Ester Gaya at Kew both indulged my request to see the fungarium and to better understand British mycology, and for helping improve the overall section on Kew. Thanks David MacNeal, for the room and board and advice at a key time in the process of writing this book, and for setting an example since our college days that has inspired me to aim higher and do better as a writer.

I'm so grateful for the time, insights, and color contributed to this book by Tradd and Olga Cotter, John Parker and Simmer, Jeff Ravage, Lauren Czaplicki, Patty Kaishian, Alan Rockefeller, Christian Schwarz, Bill Sheehan, Jonathon Keats, Amy Halliday, Tony Iwane, Paul Kroeger, Jeff Mello, Beata Tsosie-Peña, Joni Arends, Oona Goodman, Mercedes Perez Whitman, Andrew Carter, Eugenia Bone, Leah Penniman, Sue Van Hook, Sean Roy Parker, Candace Thompson, Ben Feinberg, Mendel Skulski, Charlie Aller, Nina O'Malley, Doğa Tekin, Phillip Balke, Oona Goodman, Ja Schindler and Val Nguyen at Fungi for the People, Michael Nail at Mile High Fungi, Brian Callow at WTFungus, Willy Crosby at Fungi Ally, Matt McInnis at North Spore, Daniel Reyes of the North Texas Mycological Society, and too many others to list who have informed and inspired me in ways big and small as I sought to deepen my understanding of fungi and the communities and issues to which they connect.

This book was only possible thanks to the indulgence of dear friends like John Mertens and Anne Marie; thank you for offering your home and garage at various points as I undertook an insane travel schedule. Liza Faktor, thank you for your ongoing encouragement, and for a beautiful place to live in Portland while attempting a self-imposed residency that became an unexpected quarantine shelter. Jordan Techer and Shimrit Lee, thanks for indulging my ongoing fungal soliloquies;

I'll take the oyster mushrooms that you grew in your kitchen as a sign that they were well received. Rachel Brazie, Tasia Jelatis–Hoeke, Teresa Garcia, Jake Lummus, Andrew Bailie and Katie Longofono for being my cherished chosen family; I love you all, thanks for not holding my long spans of hermitage against me. Adam and Jennifer Stacey, Judy Bankman, Andrea Balestra, David Sais, Tomas Garduno, Grayson Earle, Kyle Depew, Nisse Greenberg, Julie Gaynin, Emily Allyn, Sorya Nguyen, Gabrielle Lea, Van Bettauer, Aakash Davda, Jessica Turner, Phyllis Ma, Jie Jin, TingTing Wei, Dale Moore, John Michelotti, Luke Sarrantonio, and too many others to name, for being an essential part of my mycelial network.

Endless gratitude to my dad, Gary, whose unconditional support I've relied upon over this last year (and my entire life), as well as to my sister, Melissa, both of whom accepted and embraced my strange preoccupation about mushrooms with grace and patience and not too many raised eyebrows. Love you both.

This book was written during a time of immense transformation, for the world at large but also for myself and my family. By the end of the process, we had lost my mother, who I regret will not get to see the project I was so excited to undertake, and for which she was so excited on my behalf. Naturally, I hope that everybody who reads this book enjoys it and gets something out of it. But really, it is for her.

Notes

Introduction

1. Kent Gardner, "Rochester's Languishing Economy," *Rochester Beacon*, October 18, 2019, https://rochesterbeacon.com/2019/10/18/rochesters-languishing-economy.

2. The Mycelium Underground (website), accessed September 19, 2020, https://www.themyceliumunderground.com/.

3. Many mycophiles increasingly prefer to use the word *queendom*; while I support the spirit of this emerging convention, in the interest of being more easily legible to a variety of audiences, I will generally use the more formal Linnaean term.

4. Sevki Hakan Eren et al., "Mushroom Poisoning: Retrospective Analysis of 294 Cases," *Clinics* 65, no. 5 (May 2010): 491–96, https://dx.doi.org/10.1590%2FS1807-59322010000500006.

5. Akhil Sharma, "'If You Are Normal, You Search for Mushrooms,'" *New York Times*, October 3, 2013, https://www.nytimes.com/2013/10/06/travel/if-you-are-normal-you-search-for-mushrooms.html.

6. Kathy J. Willis, ed., *State of the World's Fungi 2018* (Royal Botanic Gardens, Kew, 2018), 9.

7. Falih Hassan and Elian Peltier, "Scorching Temperatures Bake Middle East Amid Eid al-Adha Celebrations," *New York Times*, July 31, 2020, https://www.nytimes.com/2020/07/31/world/middleeast/Middle-East-heat-wave.html.

Chapter 1: Among Us

1. Sixty percent of the enzymes used in industry come from fungi, 70 percent of which come from just seven species; see Willis, *State of the World's Fungi 2018*.

2. Elaine R. Ingham, "Soil Fungi," USDA Natural Resources Conservation Service, accessed June 25, 2020, https://www.nrcs.usda.gov/wps/portal/nrcs/detailfull/soils/health/biology/?cid=nrcs142p2_053864.

3. Peter McCoy, *Radical Mycology: A Treatise on Seeing and Working with Fungi* (Portland, OR: Chthaeus Press, 2016), 53–55.

4. Tradd Cotter, *Organic Mushroom Farming and Mycoremediation: Simple to Advanced and Experimental Techniques for Indoor and Outdoor Cultivation* (White River Junction, VT: Chelsea Green, 2014), 3.

5. James Gallagher, "More Than Half Your Body Is Not Human," *BBC News*, April 10, 2018, https://www.bbc.com/news/health-43674270.

6. "NIH Human Microbiome Project Defines Normal Bacterial Makeup of the Body," NIH News Release, June 13, 2012, https://www.nih.gov/news-events/news -releases/nih-human-microbiome-project-defines-normal-bacterial-makeup-body.

7. From a genetic perspective, the twenty thousand or so genes at the heart of our cells share our bodies with between two and twenty million microbial genes. Humans even share a fair amount of genetic code with fungi, perhaps due in part to our common heritage. In 2015, researchers at the University of Texas at Austin tested a yeast that could survive after any of its 176 genes was replaced with a human analogue. For an account of that discovery, see Marc Airhart, "Partly Human Yeast Show a Common Ancestor's Lasting Legacy," *UT Research Showcase*, The University of Texas at Austin, May 21, 2015, https://research.utexas.edu/showcase /articles/view/partly-human-yeast-show-a-common-ancestors-lasting-legacy.

8. Gallagher, "More Than Half Your Body."

9. Glenn Cardwell et al., "A Review of Mushrooms as a Potential Source of Dietary Vitamin D," *Nutrients* 10, no. 10, (October 2018): 1498, https://dx.doi.org /10.3390%2Fnu10101498.

10. Jason Arunn Murugesu, "The Oldest Fungi Fossils Have Been Identified in a Belgian Museum," *New Scientist*, January 22, 2020, https://www.newscientist. com/article/2231068-the-oldest-fungi-fossils-have-been-identified-in-a -belgian-museum.

11. Stefan Bengtson, "Fungus-Like Organisms in Deep Time and Deep Rock," Nature Research Ecology & Evolution Community, April 24, 2017, https:// natureecoevocommunity.nature.com/users/38194-stefan-bengtson /posts/16369-fungus-like-organisms-in-deep-time-and-deep-rock.

12. David Moore, Geoffrey D. Robson, and Anthony P. J. Trinci, "The Fungal Phylogeny," *21st Century Guidebook to Fungi*, 2nd ed., January 2020, http:// www.davidmoore.org.uk/21st_Century_Guidebook_to_Fungi_PLATINUM /Ch02_08.htm.

13. Bin Wang et al., "Presence of Three Mycorrhizal Genes in the Common Ancestor of Land Plants Suggests a Key Role of Mycorrhizas in the Coloniza- tion of Land by Plants," *New Phytologist* 186, no. 2 (2010): 514–25, https://doi .org/10.1111/j.1469-8137.2009.03137.x.

14. David Moore, *Fungal Biology in the Origin and Emergence of Life* (Cambridge: Cambridge University Press, 2013), 157.

15. Yinon M. Bar-On, Rob Phillips, and Ron Milo, "The Biomass Distribution on Earth," *Proceedings of the National Academy of Sciences* 115, no. 25 (June 2018): 6506–11, Fig. 1, https://doi.org/10.1073/pnas.1711842115.

16. Ekta Khare, Jitendra Mishra, and Naveen Kumar Arora, "Multifaceted Interac- tions Between Endophytes and Plant: Developments and Prospects," *Frontiers in Microbiology* 15 (November 2018), https://doi.org/10.3389/fmicb.2018.02732;

Kusam Lata Rana et al., "Endophytic Fungi: Biodiversity, Ecological Signifi-
cance, and Potential Industrial Applications," in *Recent Advancement in White
Biotechnology Through Fungi*, vol. 1, *Diversity and Enzymes Perspectives* (Cham,
CH: Springer, 2019), 1–62, https://doi.org/10.1007/978-3-030-10480-1_1.

17. Michael Phillips, *Mycorrhizal Planet* (White River Junction, VT: Chelsea
Green, 2017), 6.

18. By comparing the fungus's size to its rate of spread—about one to three feet a
year—scientists estimated it to be between 1,900 and 8,650 years old; see Craig
L. Schmitt and Michael L. Tatum, "The Malheur National Forest Location of
the World's Largest Living Organism [The Humongous Fungus]," United States
Department of Agriculture, Forest Service, Pacific Northwest Region (2008), 4,
https://www.fs.usda.gov/Internet/FSE_DOCUMENTS/fsbdev3_033146.pdf.

19. Gero Steinberg, "Hyphal Growth: A Tale of Motors, Lipids, and the Spitzen-
körper," *Eukaryotic Cell* 6, no. 3 (March 2007): 351–60, https://doi.org
/10.1128/ec.00381-06.

20. Cotter, *Organic Mushroom Farming*, 7–8.

21. McCoy, *Radical Mycology*, 14.

22. Phillips, *Mycorrhizal Planet*, 6.

23. Fungi can also parasitize other fungi. Perhaps the most often cited example
is the delectable lobster mushroom, which develops a signature red "shell"
on its surface when *Hypomyces lactifluorum* parasitizes a species of *Russula*
or *Lactarius*, turning it from bland or overly spicy to crunchy and tasty; see
Tom Volk, "This Month's Fungus Is *Hypomyces lactifluorum*, the Lobster
Mushroom," *Tom Volk's Fungus of the Month for August, 2001*, accessed August
6, 2020, https://botit.botany.wisc.edu/toms_fungi/aug2001.html.

24. *Phallus impudicus*, for example, is a slender, pale, sometimes slightly curved
mushroom with a dark, rounded tip that bears an undeniable resemblance to,
well, you get the idea.

25. Paul Money, "More G's Than the Space Shuttle: Ballistospore Discharge," *Mycologia*
90, no. 4 (July 1998): 547–58, https://doi.org/10.1080/00275514.1998.12026942.

26. Emilie Dressaire et al., "Mushrooms Use Convectively Created Airflows to
Disperse Their Spores," *Proceedings of the National Academy of Sciences* 113, no.
11 (March 2016): 2833–38, https://doi.org/10.1073/pnas.1509612113.

27. Athanasios Damialis et al., "Estimating the Abundance of Airborne Pollen and
Fungal Spores at Variable Elevations Using an Aircraft: How High Can They
Fly?" *Scientific Reports* 7 (March 16, 2017): article 44535, https://doi.org
/10.1038/srep44535.

28. Amy M. Hanson, Kathie T. Hodge, and Leila M. Porter, "Mycophagy Among
Primates," *Mycologist* 17, no. 1 (February 2003): 6–10, https://doi.org/10.1017
/S0269915X0300106X.

29. Daniel B. Raudabaugh et al., "Where Are They Hiding? Testing the Body Snatchers Hypothesis in Pyrophilous Fungi," *Fungal Ecology* 43 (October 2019): 1–10, https://doi.org/10.1016/j.funeco.2019.100870.
30. Phillips, *Mycorrhizal Planet*, 66.
31. Nicholas P. Money, *Mr. Bloomfield's Orchard* (New York: Oxford University Press, 2002), 58.
32. Donald P. Rogers, "Basidial Proliferation Through Clamp-Formation in a New Sebacina," *Mycologia* 28, no. 4 (1936): 347–62, https://doi.org/10.1080/00275514.1936.12017150.
33. I have since been informed that the *Amanita muscaria* and *Boletus edulis* tend to fruit at the same time.
34. Unprepared as I was, the bag I brought was plastic, which I soon understood to be a no-no among mushroom pickers. Besides being environmental pollutants, plastic bags trap moisture and hasten the decomposition of the mushrooms. Pickers prefer paper or, ideally, cloth bags, so that the spores can drift and spread as they walk.
35. Lewis Carroll, *Alice's Adventures in Wonderland* (Boston: Lee and Shepard, 1869), 68.
36. John A. Rush, *Entheogens and the Development of Culture* (Berkeley, CA: North Atlantic Books, 2013), 283.
37. Rush, *Entheogens*, 284–86.
38. Rush, 292.
39. Matthew Salton, "Santa Is a Psychedelic Mushroom," *New York Times*, December 21, 2017, https://www.nytimes.com/2017/12/21/opinion/santa-christmas-mushrooms.html.
40. Andy Letcher, *Shroom: A Culural History of the Magic Mushroom* (New York: HarperCollins, 2007), 137–38.
41. Simon Critchley, "Athens in Pieces: What Really Happened at Eleusis?," *New York Times*, March 13, 2019, https://www.nytimes.com/2019/03/13/opinion/ancient-greece-ritual-mystery-eleusis.html.
42. Mara Lynn Keller, PhD, "The Ritual Path of Initiation into the Eleusinian Mysteries," *Rosicrucian Digest* 90, no. 2 (2009): 28–42.
43. EcoFarmVideo, "Mushrooms, Mycology of Consciousness—Paul Stamets, EcoFarm Conference Keynote 2017," YouTube Video, 1:20:53, February 16, 2017, https://www.youtube.com/watch?v=XI5frPV58tY.
44. Letcher, *Shroom*, 28–29.
45. Linnda R. Caporael, "Ergotism: The Satan Loosed in Salem?" *Science* 192, no. 4234 (April 2, 1976): 21–26, https://doi.org/10.1126/science.769159.
46. Ursula Peintner, Reinhold Pöder, and Thomas Pümpel, "The Iceman's Fungi," *Mycological Research* 102, no. 10 (October 1, 1998): 1153–62, https://doi.org/10.1017/S0953756298006546.

47. Robert C. Power et al., "Microremains from El Mirón Cave Human Dental Calculus Suggest a Mixed Plant Animal Subsistence Economy During the Magdalenian in Northern Iberia," *Journal of Archaeological Science* 60 (August 2015): 39–46, https://doi.org/10.1016/j.jas.2015.04.003.
48. The mushroom is assessed as "near threatened" on the IUCN Red List as of June 2020.
49. Robert A. Blanchette, "Haploporus Odorus: A Sacred Fungus in Traditional Native American Culture of the Northern Plains," *Mycologia* 89, no. 2 (March–April 1997): 233–40, https://doi.org/10.2307/3761076.
50. Diane Pleninger and Tom Volk, "Phellinus igniarius, Iqmik, Used by Native Americans with Tobacco," *Tom Volk's Fungus of the Month for November 2005*, accessed June 25, 2020, https://botit.botany.wisc.edu/toms_fungi/nov2005.html.
51. George W. Hudler, *Magical Mushrooms, Mischievous Molds* (Princeton, NJ: Princeton University Press, 1998), 124–26.
52. Aamir Peerzada, "The Perilous Search for 'Himalayan Viagra,'" *BBC News*, July 17, 2018, https://www.bbc.com/news/av/business-44846135/the-perilous -search-for-himalayan-viagra.
53. Greg A. Marley, *Chanterelle Dreams, Amanita Nightmares* (White River Junction, VT: Chelsea Green, 2010), 17.
54. Charles McIlvaine and Robert K. Macadam, preface to *Toadstools, Mushrooms, Fungi, Edible and Poisonous: One Thousand American Fungi: How to Select and Cook the Edible; How to Distinguish and Avoid the Poisonous* (Indianapois, MD: The Bobbs-Merrill Company, 1912), viii.
55. There is, incidentally, no mushroom that it is unsafe to touch, unless it's one that has been eyed by a particularly jealous or overzealous picker.
56. Fundación Fungi, "Our Achievements," 2019, https://ffungi.org/eng/our-impact.
57. Willis, ed., *State of the World's Fungi*, 77.
58. "Aprueba Reglamento del Sistema de Evaluación de Impacto Ambiental," *Biblioteca del Congresso Nacional de Chile*, August 12, 2010, http://bcn.cl/1uvqa.
59. Giuliana Furci, *Guias de Campo Hongos de Chile*, vol. 2 (Santiago, CL: Fundación Fungi, 2018).

Chapter 2: Scratching the Surface

1. Julie Robinson and Kirt Costello, eds., *International Space Station Benefits for Humanity*, 3rd ed. (International Space Station Program Science Forum, 2019), 159.
2. David L. Hawksworth, "Lichenization: The Origins of a Fungal Life-Style," *Recent Advances in Lichenology: Modern Methods and Approaches in Lichen Systematics and Culture Techniques* 2 (2015): 1–10, https://doi.org/10.1007 /978-81-322-2235-4_1.

3. *Arbus* being the root word, as it were, for "tree."

4. Jonathan Leake et al., "Networks of Power and Influence: The Role of Mycorrhizal Mycelium in Controlling Plant Communities and Agroecosystem Functioning," *Canadian Journal of Botany* 82, no. 8 (2004): 1016–45, https://doi.org/10.1139/b04-060.

5. Winxiao Wang et al., "Nutrient Exchange and Regulation in Arbuscular Mycorrhizal Symbiosis," *Molecular Plant* 10, no. 9 (September 12, 2017): 1147–58, https://doi.org/10.1016/j.molp.2017.07.012.

6. Phillips, *Mycorrhizal Planet*, 62.

7. Zhi-Gang Wang et al., "Arbuscular Mycorrhizal Fungi Enhance Soil Carbon Sequestration in the Coalfields, Northwest China," *Scientific Reports* 6, no. 34336 (2016): https://doi.org/10.1038/srep34336.

8. Regina S. Redman et al., "Fungal Symbiosis from Mutualism to Parasitism: Who Controls the Outcome, Host or Invader?" *New Phycologist* 151, no. 3 (December 21, 2001), https://doi.org/10.1046/j.0028-646x.2001.00210.x.

9. Paul Stamets, "6 Ways Mushrooms Can Save the World," TED, video, 18:17, May 8, 2008, https://www.youtube.com/watch?v=XI5frPV58tY.

10. David C. Coleman et al., "Primary Production Processes in Soils: Roots and Rhizosphere Associates," *Fundamentals of Soil Ecology*, 2nd ed. (2004), 23–46, https://doi.org/10.1016/B978-012179726-3/50003-4.

11. Phillips, *Mycorrhizal Planet*, 12.

12. Phillips, *Mycorrhizal Planet*, 14.

13. Suzanne W. Simard, "Mycorrhizal Networks Facilitate Tree Communication, Learning, and Memory," *Memory and Learning in Plants* (2018): 191–213, https://doi.org/10.1007/978-3-319-75596-0_10.

14. Paul Stamets, *Mycelium Running: How Mushrooms Can Help Save the World* (Berkeley, CA: Ten Speed Press, 2005), 73–74.

15. Kristine Nichols, "Glomalin: Hiding Place for a Third of the World's Stored Soil Carbon," *Agricultural Research* 50, no. 9 (September 2002): 4–7.

16. Jennifer Frazer, "Root Fungi Can Turn Pine Trees into Carnivores—or at Least Accomplices," *Scientific American*, May 12, 2015, https://blogs.scientific american.com/artful-amoeba/root-fungi-can-turn-pine-trees-into-carnivores -8212-or-at-least-accomplices.

17. Chakravarthulu Manoharachary et al., "Some Aspects of Monotropoid Mycorrhizas," *Techniques in Mycorrhizal Studies* (2002): 435–41, https://doi .org/10.1007/978-94-017-3209-3_22.

18. Marcin Nowicki et al., "Potato and Tomato Late Blight Caused by *Phytophthora infestans*: An Overview of Pathology and Resistance Breeding," *American Phytopathological Society* 96, no. 1 (December 13, 2011): 4–17, https://doi .org/10.1094/PDIS-05-11-0458.

19. Cheryl Katz, "Small Pests, Big Problems: The Global Spread of Bark Beetles," *Yale Environment 360*, September 21, 2017, https://e360.yale.edu/features /small-pests-big-problems-the-global-spread-of-bark-beetles.

20. "Areas with Tree Mortality from Bark Beetles, Summary for 2000–2019 Western US," (USDA/US Forest Service, March 20, 2020), 1, https://www .fs.fed.us/foresthealth/applied-sciences/news/2019/index.shtml.

21. Staff, "Important Food Source for Yellowstone Bears in Trouble," *National Park Trips Media*, September 19, 2019, https://www.yellowstonepark.com/park /grizzly-food-source.

22. Forest Staff, "Don't Trip Over Your Shoestrings! Shoestring Root Rot, That Is!" *In the Field*, October 11, 2017, http://info.ncagr.gov/blog/2017/10/11 /dont-trip-over-your-shoestrings-shoestring-root-rot-that-is/.

23. Ryan S. Davis, "Elm Bark Beetles and Dutch Elm Disease," Utah State University Extension and Utah Plant Pest Diagnostic Laboratory, September 2011, https://digitalcommons.usu.edu/cgi/viewcontent.cgi?article=1895 &context=extension_curall.

24. Diana L. Six and James J. Elser, "Extreme Ecological Stoichiometry of a Bark Beetle–Fungus Mutualism," *Ecological Entomology* (2019): 2, https://doi.org /10.1111/een.12731.

25. Brian T. Sullivan, "Semiochemicals in the Natural History of Southern Pine Beetle Dendroctonus Frontalis Zimmermann and Their Role in Pest Management," chap. 4 in *Advances in Insect Physiology* 50 (2016): 129–93, https://doi .org/10.1016/bs.aiip.2015.12.002.

26. Diana L. Six and Michael J. Wingfield, "The Role of Phytopathogenicity in Bark Beetle–Fungus Symbioses: A Challenge to the Classic Paradigm," *Annual Review of Entomology* 56 (2011): 255–72, https://doi.org/10.1146/annurev -ento-120709-144839.

27. Six and Wingfield, "The Role of Phytopathogenicity," 261.

28. Six and Wingfield, 258.

29. Even though the majority of beetle species are believed to have fungal associations, a minority—less than fifteen of the thousands of bark beetle species worldwide—are actually known to be deadly to trees; see Six and Wingfield, 263.

30. Leia Larsen, "Bark Beetle Research Shows Future Evolution of Utah Forests," *Standard-Examiner*, October 7, 2014, Environment, https://www.standard .net/news/environment/bark-beetle-research-shows-future-evolution-of-utah -forests/article_1cd73a18-09ad-519a-a1f5-d5ede356189e.html.

31. Maddie Oatman, "Bark Beetles Are Decimating Our Forests. That Might Actually Be a Good Thing," *Mother Jones* (May/June 2015), https://www.motherjones .com/environment/2015/03/bark-pine-beetles-climate-change-diana-six.

32. Cameron R. Currie, "The Tangled Banks of Ants and Microbes," *Microbes and Evolution* (2014): 181–90, https://doi.org/10.1128/9781555818470.ch25.

Chapter 3: A Neglected Megascience

1. Willis, ed., *State of the World's Fungi*, 9.
2. Jack-o'-lantern mushrooms, for example, are sometimes mistaken for chanterelles. One interesting trait that sets them apart is that the former glow in the dark, presumably as an insect attractant, thanks to the evocatively named compound luciferin, the same type of chemical responsible for bioluminescence in fireflies.
3. Michael Kuo, "Mushrooming in the Age of DNA: Now Comes the Fun Part," *McIlvainea* 17, no. 1 (Spring 2007): 43–49, https://namyco.org/docs/DNA.pdf.
4. Ali R. Awan et al., "Convergent Evolution of Psilocybin Biosynthesis by Psychedelic Mushrooms," *BioRxiv 374199* (July 27, 2018), https://doi.org/10.1101/374199.
5. Kennedy Warne, "Organization Man," *Smithsonian Magazine*, May 2007, https://www.smithsonianmag.com/science-nature/organization-man-151908042.
6. John G. Meert., "Gondwanaland, Formation," in *Encyclopedia of Geobiology*, Encyclopedia of Earth Sciences Series (Dordrecht, ND: Springer, 2011), https://doi.org/10.1007/978-1-4020-9212-1_92.
7. Marc D. Meyer et al., "Fungi in the Diets of Northern Flying Squirrels and Lodgepole Chipmunks in the Sierra Nevada," *Canadian Journal of Zoology* 83 (2005): 1582–88, https://www.fs.fed.us/psw/publications/meyer/captured/psw_2005_meyer003.pdf.
8. Robert H. Whittaker, "New Concepts of Kingdoms of Organisms. Evolutionary Relations Are Better Represented by New Classifications Than by the Traditional Two Kingdoms," *Science, New Series* 163, no. 3863 (January 10, 1969): 150–60, https://doi.org/10.1126/science.163.3863.150.
9. Alfred Lambourne, *Pictures of an Inland Sea: The Story of a Homestead* (Salt Lake City: The Deseret News, 1902), 47.
10. Bonnie K. Baxter and Polona Zalar, "The Extremophiles of Great Salt Lake: Complex Microbiology in a Dynamic Hypersaline Ecosystem," *Model Ecosystems in Extreme Environments, Astrobiology Exploring Life on Earth and Beyond* (2019): 57–99, https://doi.org/10.1016/B978-0-12-812742-1.00004-0.
11. John Luft, "Facts," Great Salt Lake Ecosystem Project, https://wildlife.utah.gov/gsl/facts/properties.php; Swati Almeida-Dalmet and Bonnie K. Baxter, "Unexpected Complexity at Salinity Saturation: Microbial Diversity of the North Arm of Great Salt Lake," *Great Salt Lake Biology* (2020): 119–44, https://doi.org/10.1007/978-3-030-40352-2_5.
12. US Geological Survey, "Great Salt Lake, Utah," *Water-Resources Investigations Report*, April 2007, https://pubs.usgs.gov/wri/wri994189/PDF/WRI99-4189.pdf.

13. Baxter and Zalar, "Extremophiles," 73.

14. Nine Gunde-Cimerman et al., eds., introduction to *Adaptation to Life at High Salt Concentrations in Archaea, Bacteria, and Eukarya* (Dordrecht, ND: Springer, 2005), 4.

15. Bonnie K. Baxter, "Great Salt Lake Microbiology: A Historical Perspective," *International Microbiology* 21 (June 4, 2018): 79–95, https://doi.org/10.1007/s10123-018-0008-z.

16. Don S. Paul and Ann E. Manning, "Great Salt Lake Waterbird Survey Five-Year Report 1997–2001," (State of Utah Department of Natural Resources, Division of Wildlife Resources, Great Salt Lake Ecosystem Program (December 2002), 38, https://wildlife.utah.gov/gsl/gsl_ws_report/gsl_ws_report.pdf.

17. Leia Larsen, "Mineral Extraction on Great Salt Lake Has Local, National and Global Impact," *Standard-Examiner*, May 1, 2016, https://www.standard.net/news/environment/mineral-extraction-on-great-salt-lake-has-local-national-and/article_875d73e9-9100-54a5-a0e9-371e012c3945.html.

18. Baxter, "Great Salt Lake Microbiology," 83.

19. Douglas Fox, "Utah's Ancient Lake Bonneville Holds Clues to the West's Changing Climate," *High Country News*, November 7, 2011, https://www.hcn.org/issues/43.18/utahs-ancient-lake-bonneville-holds-clues-to-the-wests-changing-climate.

20. Baxter, "Great Salt Lake Microbiology," 82.

21. Forrest S. Cuch, ed., "History of Utah's American Indians" (Salt Lake City: Utah State Division of Indian Affairs and the Utah State Division of History, 2003), 4.

22. Baxter, "Great Salt Lake Microbiology," 82.

23. Richard E. Fike and John W. Headley, introduction to "The Pony Express Stations of Utah in Historical Perspective," *Bureau of Land Management, Utah, Cultural Resources Series*, monograph 2 (1979).

24. Dennis R. Defa, "The Goshute Indians of Utah," in *History of Utah's American Indians* (Louisville: University Press of Colorado, 2000), 104.

25. Ben Pierce, "Jim Bridger: The Man, the Myth, the Legend," *Seattle Times*, February 6, 2016, https://www.seattletimes.com/nation-world/jim-bridger-the-man-the-myth-the-legend/.

26. Edward Young, "Night-Thoughts," quoted in Howard Stansbury, "Exploration of the Valley of the Great Salt Lake: Including a Reconnaissance of a New Route Through the Rocky Mountains," (Philadelphia: Lippincott, Gramabo & Co., 1855), 197.

27. Baxter, "Great Salt Lake Microbiology," 83–84.

28. Baxter, 79.

29. Tim Brady, "Of Algae and Acrimony," *Minnesota Magazine* (January–February 2008), 30–33.

30. Baxter, "Great Salt Lake Microbiology," 82.

31. Brady, "Of Algae and Acrimony," 31.

32. Baxter, "Great Salt Lake Microbiology," 86.
33. Baxter, 89.
34. Baxter, 88–89.
35. Baxter, 87.
36. Asya S. Buchalo et al., "Fungal Life in the Extremely Hypersaline Water of the Dead Sea: First Records," *The Royal Society, Procession of the Natural Sciences* 265, no. 1404 (August 7, 1998): 1461–65, https://dx.doi.org/10.1098%2Frspb.1998.0458.
37. Sharon A. Cantrell et al., "Characterization of Fungi from Hypersaline Environments of Solar Salterns Using Morphological and Molecular Techniques," *Mycological Research* 110, no. 8 (August 10, 2006): 962–70, https://doi.org/10.1016/j.mycres.2006.06.005.
38. In fact, the very PCR techniques used by labs like those at NHMU were made possible only by the discovery of an extremophilic bacteria. In 1969, biologist Thomas D. Brock identified the bacteria *Thermus aquaticus*, which ultimately earned him a Nobel Prize; the bacteria he discovered were later found by biochemist Kary Mullis to contain a special enzyme that can withstand high temperatures, making it useful in the DNA copying process. Those bacteria were first isolated in, of all places, Mushroom Spring at Yellowstone National Park.
39. Baxter and Zalar, "Extremophiles," 73.
40. Letcher, *Shroom*, 62–63.
41. Emory G. Simmons, "The International Mycological Association: Its History in Brief with Summaries of Its International Mycological Congresses and Diverse International Relationships," *IMA Fungus* 1, no. 1 (June 2010): 18–100, https://dx.doi.org/10.5598%2Fimafungus.2010.01.01.01.
42. Jeffrey Mervis, "Trump's New Budget Cuts All But a Favored Few Science Programs," *Science* (February 11, 2020), https://www.sciencemag.org/news/2020/02/trump-s-new-budget-cuts-all-favored-few-science-programs.

Chapter 4: Big *S*, Small *S*

1. Kew Botanic Gardens, "Our Collections," accessed August 5, 2020, https://www.kew.org/science/collections-and-resources/collections.
2. Kew Botanic Gardens, "The Fungarium," accessed August 5, 2020, https://www.kew.org/science/collections-and-resources/collections/fungarium.
3. Adam Gabbat, "Marmite: Americans Wonder What's All the Fuss over Divisive British Spread?" *The Guardian*, October 13, 2016, https://www.theguardian.com/lifeandstyle/2016/oct/13/what-is-marmite-british-food-spread-tesco.
4. Alexandre Antonelli, "The Time Has Come to Decolonise Botanical Gardens Like Kew," *The Independent*, June 26, 2020, https://www.independent.co.uk/news/uk/kew-gardens-black-lives-matter-decolonise-botanical-a9585661.html.

5. "Summary Statistics," IUCN Red List (website), accessed August 5, 2020, https://www.iucnredlist.org/statistics.

6. Rebecca Morelle, "Kew Gardens Funding Cuts 'Recipe for Failure,'" *BBC News*, March 4, 2015, https://www.bbc.com/news/science-environment-31715081.

7. Dr. Oliver Ellingham et al., "The Final Year of the Lost and Found Fungi Project," March 28, 2018, https://www.kew.org/read-and-watch/final-year-lost-and-found.

8. Dr. Brian Douglas and Dr. Oliver Ellingham, "Lost and Found Fungi Project —Statistics," February 2019, http://fungi.myspecies.info/content/lost-and -found-fungi-project-statistics.

9. "Historique," MycoFrance, accessed June 23, 2020, http://www.mycofrance.fr /smf/historique/.

10. "Recording Fungi for Local Citizen Science," Bento Labs, video, 2:29, March 24, 2016, https://www.youtube.com/watch?v=9DaUKTt3FEQ.

11. "A Foray Among the Funguses," *Transactions of the Woolhope Naturalists' Field Club* (Hereford, UK: October 9, 1868), 184–92, https://www.biodiversity library.org/item/44662#page/184/mode/1up.

12. John Webster, "The British Mycological Society, 1896–1996," *Mycological Research* 101, no. 10 (October 1997), 1153–78, https://doi.org/10.1017 /S0953756297004553.

13. John Webster and David Moore, eds., introduction to *Brief Biographies of British Mycologists*, by Geoffrey C. Ainsworth (Stourbridge, UK: British Mycological Society, 1996), v.

14. Olivia Campbell, "Under Victorian Microscopes, an Enchanted World," *JSTOR Daily* (February 21, 2018), https://daily.jstor.org/victorian-microscope -enchanted-world/.

15. Will Kaufman and Heidi Slettedahl Macpherson, eds., *Britain and the Americas: Culture, Politics, and History, a Multidisciplinary Encyclopedia*, vol. 1 (Santa Barbara, CA: ABC–CLIO, 2005), 869–76.

16. The Royal Society, "History of the Royal Society," accessed June 26, 2020, https://royalsociety.org/about-us/history/.

17. The Linnean Society of London, "Communicating Nature Since 1788," accessed June 26, 2020, https://www.linnean.org/the-society.

18. Anna Agnarsdóttir, ed., *Sir Joseph Banks, Iceland and the North Atlantic, 1772–1820: Journals, Letters and Documents* (Oxon, UK: Routledge, 2016), 25–29.

19. Ainsworth, *Brief Biographies*, 13–15.

20. Ainsworth, 43–45.

21. Letcher, *Shroom*, 63–65.

22. Ainsworth, *Brief Biographies*, 113–15.

23. Matthew Wills, "The Other Side of Beatrix Potter," *JSTOR Daily*, March 7, 2018, https://daily.jstor.org/the-other-side-of-beatrix-potter/.

24. Kiri Ross-Jones, "Beatrix Potter: Tales from the Archives," Kew Botanic Gardens, March 29, 2018, https://www.kew.org/read-and-watch/beatrix -potter-tales-from-the-archives.

25. McCoy, *Radical Mycology*, 82–84.

26. R.W.G. Dennis and E.M. Wakefield, "New or Interesting British Fungi," *Transactions of the British Mycological Society* 29, no. 3, (September 1946), 141–66, https://doi.org/10.1016/S0007-1536(46)80038-X.

27. Jed Lipinski, "On Flatbush Avenue, Seven Stories Full of Ideas," *New York Times*, January 11, 2011, https://www.nytimes.com/2011/01/12/realestate /commercial/12incubate.html.

28. McCoy, *Radical Mycology*, 393.

29. Jan Borovička et al., "Psilocybe Allenii—A New Bluing Species from the Pacific Coast, USA," *Czech Mycology* 64, no. 2 (2012): 181–95.

30. [Amadeus Grows], "Alan Rockefeller Should Partner with Vice for a Series About Worldwide Fungi," Change.org, accessed June 25, 2020, https://www .change.org/p/vice-alan-rockefeller-should-partner-with-vice-for-a-series -about-worldwide-fungi.

31. Jennifer McNulty, "Alumni Profile: Author-Naturalist Christian Schwarz Crusades on Behalf of the Planet," *UC Santa Cruz News Center*, October 4, 2018, https://news.ucsc.edu/2018/10/schwarz-fungi.html.

32. A mycoflora is the fungal version of a flora, or a list of the plant and fungal life of a particular region or period; dissatisfaction with the term is part of what led the North American Mycoflora Project to change its name. The term *funga* is sometimes used as an alternative.

33. Sergi Santamaria et al., "The First Laboulbeniales (Ascomycota, Laboulbeniomy-cetes) from an American Millipede, Discovered Through Social Media," *MycoKeys* 67 (May 14, 2020): 45–53, https://doi.org/10.3897/mycokeys.67.51811.

34. Nathan Agrin et al., "iNaturalist.org Final Project Write-Up," (master's final project report, UC Berkeley, 2008), https://www.ischool.berkeley.edu/sites /default/files/iNaturalist_Final_Writeup.pdf.

35. Scott Loarie, "25,000,000 Observations!" iNaturalist Blog, August 15, 2019, https://www.inaturalist.org/blog/26737-25-000-000-observations.

36. Tony Iwane, "A Growing iNaturalist Community in Mumbai Documents Their Beach Finds," iNaturalist Blog, September 18, 2018, https://www.inaturalist .org/blog/18850-a-growing-inaturalist-community-in-mumbai-documents -their-beach-finds.

37. *Macrofungi* is another term for "mushroom," referring to fungi that express themselves in a visible form.

38. "FunDiS Project Map," FunDis, accessed August 8, 2020, https://fundis.org /about/project-map.

39. Michael J. O. Pocock et al., "The Diversity and Evolution of Ecological and Environmental Citizen Science," *PLoS ONE* 12, no. 4 (April 3, 2017), https://doi.org/10.1371/journal.pone.0172579.

40. Bill Sheehan, "A Four-Tiered Model for Crowdsourcing Fungal Biodiversity Citizen Science," *The Startup* (February 22, 2020), https://medium.com/swlh/a-four-tiered-model-for-crowdsourcing-fungal-biodiversity-citizen-science-b8928cec2209.

41. George F. Barrowclough et al., "How Many Kinds of Birds Are There and Why Does It Matter?" *PLoS ONE* 11, no. 11 (November 23, 2016), https://doi.org/10.1371/journal.pone.0166307.

42. Willis, ed., *State of the World's Fungi*, 9.

43. Team eBird, "eBird 2019—Year in Review," *eBird News* (December 23, 2019), https://ebird.org/news/ebird-2019-year-in-review.

44. Stephen Russell [stevilkinevil], "Continental Mycoblitz 2019," iNaturalist (project page), August 11–October 27, 2019, https://www.inaturalist.org/projects/continental-mycoblitz-2019.

Chapter 5: This Land Is Mycoland

1. Amy M. Peters, "Mushroom Fest Returns This Weekend" (August 18, 2016), *Telluride Daily Planet*, https://www.telluridenews.com/news/article_f39225d4-659b-11e6-9f28-733dfd85ed93.html.

2. Panspermia suggests that life on Earth was seeded from outer space. Those who believe in it say that fungal spores, tough as they are, fell from the sky from outer space. There is good reason to question this account, as discussed in Moore, *Fungal Biology*, 62–84.

3. Ella Riley-Adams and Caroline Tompkins, "The Place Where Mushrooms Get Their Own Parade," *Vogue* (September 12, 2019), https://www.vogue.com/slideshow/telluride-mushroom-festival-2019.

4. Michael Pollan, *How to Change Your Mind* (New York: Penguin Press, 2018), 19.

5. The term *psychedelic* was coined by British psychiatrist Humphry Osmond, chosen to replace the term *psychotomimetic*, as the field of psychiatry carried a certain social stigma at the time. The term also won out over an alternative proposed by Aldous Huxley: *phanerothyme*. For further discussion, see Marley, *Chanterelle Dreams*, 143–47.

6. Benedict Carey, "Johns Hopkins Opens New Center for Psychedelic Research," *New York Times*, September 4, 2019, https://www.nytimes.com/2019/09/04/science/psychedelic-drugs-hopkins-depression.html.

7. The famed and revered biochemist is credited with introducing MDMA, along with hundreds of other psychoactive compounds that he synthesized and tested; see Pollan, *How to Change Your Mind*, 44.

8. The phrase was added to the *Oxford English Dictionary* in 2004, along with *larper*.

9. Subjects of psilocybin studies have reported that the experience under the influence of the compound count among the top five most powerful and positive
in test subjects' lives, compared in their profundity with marriage, the birth of
children, and the death of parents; see Roland R. Griffiths et al., "Mystical-Type
Experiences Occasioned by Psilocybin Mediate the Attribution of Personal
Meaning and Spritual Significance 14 Months Later," *Journal of Psychopharma-
cology* 22, no. 6 (2008), https://dx.doi.org/10.1177%2F0269881108094300.

10. Matt Simon, "The Heady, Thorny Journey to Decriminalize Magic Mushrooms," *Wired*, June 19, 2019, https://www.wired.com/story/the-heady
-thorny-journey-to-decriminalize-magic-mushrooms/.

11. Melanie J. Miller et al., "Chemical Evidence for the Use of Multiple Psy-
chotropic Plants in a 1,000-Year-Old Ritual Bundle from South America,"
Proceedings of the National Academy of Sciences of the United States of America 116,
no. 23 (June 4, 2019): 11207–12, https://doi.org/10.1073/pnas.1902174116.

12. Letcher, *Shroom*, 82–83.

13. Robert Gordon Wasson, "Seeking the Magic Mushroom," *Life* 49, no. 19 (May
13, 1957): 114.

14. Wasson, "Seeking the Magic Mushroom," 113.

15. Letcher, *Shroom*, 88.

16. Felipe Ruan-Soto, "Evaluation of the Degree of Mycophilia-Mycophobia
Among Highland and Lowland Inhabitants from Chiapas, Mexico," *Journal of
Ethnobiology and Ethnomedicine* 9, no. 36 (May 26, 2013), https://doi.org
/10.1186/1746-4269-9-36.

17. Marley, *Chanterelle Dreams*, 192–99.

18. William Shakespeare, *The Tempest*, ed. Samuel Thurber (Norwood, MA:
Norwood Press, 1900), 5.1.36–40.

19. Letcher, *Shroom*, 74–78.

20. Letcher, 193–94.

21. Letcher, 76–77.

22. Pollan, *How to Change Your Mind*, 22–25.

23. Pollan, 104.

24. A common feature and, arguably, distinct genre of writing found in psychedelic
culture in magazines and online forums, in which someone attempts to put to
words their ultimately indescribable experience of consuming a mind-altering
substance such as psilocybin, salvia, or ayahuasca.

25. Eric Zolov, *Refried Elvis: The Rise of the Mexican Counterculture* (Berkeley:
University of California Press, 1999), 141–46.

26. Controlled Substances Act, Title 21 of the United States Code, chapter 13,
May 1, 1971.

27. Paul Kroeger, "A Brief History of Magic Mushrooms in BC," Vancouver

Mycological Society (website), March 9, 2018, https://www.vanmyco.org
/about-mushrooms/psychedelic/brief-history-magic-mushrooms-bc/.

28. Grant Lawrence, "Positively 4th Avenue: The Rise and Fall of Canada's Hippie
Mecca," *Vancouver Courier*, October 19, 2016, https://www.vancourier.com/news
/positively-4th-avenue-the-rise-and-fall-of-canada-s-hippie-mecca-1.2368955.

29. Kroeger, "Magic Mushrooms in BC."

30. Stamets, *Mycelium Running*, 294.

31. Debbie Viess, "Amanita Phalloides: Invasion of the Death Cap," Bay Area
Mycological Society (website), accessed August 5, 2020, http://bayarea
mushrooms.org/mushroommonth/amanita_phalloides.html.

32. Craig Childs, "Death-Cap Mushrooms Are Spreading Across North America,"
The Atlantic, February 1, 2019, https://www.theatlantic.com/science/archive
/2019/02/deadly-mushroom-arrives-canada/581602/.

33. "Amanita Phalloides," Vancouver Mycological Society, accessed August 5, 2020,
https://www.vanmyco.org/about-mushrooms/poisonous/amanita-phalloides/.

34. David W. Rose, "History of NAMA," North American Mycological Society
(website), accessed June 27, 2020, https://namyco.org/history_of_nama.php.

35. Eugenia Bone, *Mycophilia: Revelations from the Weird World of Mushrooms* (New
York: Rodale, 2011), 4–7.

36. Sabrina Small, "Harmony of the Spores: John Cage and Mycology," *Gastronom-
ica: The Journal of Food and Culture* 11, no. 2 (2011), https://doi.org/10.1525
/gfc.2011.11.2.19.

37. Alleen Brown and Amber Bracken, "No Surrender," *The Intercept*, February 23, 2020,
https://theintercept.com/2020/02/23/wetsuweten-protest-coastal-gaslink-pipeline/.

38. Although he famously despises the term *shroom*; see Pollan, *How to Change Your
Mind*, 93.

39. This notion was recently given a bit of extra heft, as astronomers used slime
mold (albeit not a fungus, but often mistaken for them) to model the spread of
dark matter throughout the universe; see Joseph N. Burchett et al., "Revealing
the Dark Threads of the Cosmic Web," *Astrophysical Journal Letters* 891, no. 2
(March 10, 2020), https://doi.org/10.3847/2041-8213/ab700c.

40. Stamets, *Mycelium Running*, 114–23; Paul E. Stamets et al., "Extracts of Polypore
Mushroom Mycelia Reduce Viruses in Honey Bees," *Scientific Reports* 8, 13936
(October 4, 2018), https://www.nature.com/articles/s41598-018-32194-8.

41. Merlin Sheldrake, *Entangled Life: How Fungi Make Our Worlds, Change Our Minds,
and Shape Our Futures* (New York: Penguin Random House, 2020), 197–98.

42. Letcher, *Shroom*, 250–75.

43. Terence McKenna, "Psychedelics Before and After History," filmed 1987 at the
California Institute for Integral Studies, video, 1:58:39, https://www.youtube
.com/watch?v=hcRGY2Bdk0U.

44. Joanna Beth Steinhardt, "Mycelium Is the Message: Open Science, Ecological Values, and Alternative Futures with Do-It-Yourself Mycologists" (PhD diss., University of California, Santa Barbara, September 2018), 170.

45. Terence McKenna, foreword to *Psilocybin, Magic Mushroom Grower's Guide* (Berkeley, CA: And/Or Press, 1976), 13.

46. *Psilocybe* mushrooms are well known for bruising blue, a result of oxidation when exposed to the air.

47. Hamilton Morris, "Blood Spore," *Harper's*, July 2013, https://harpers.org /archive/2013/07/blood-spore/.

48. Like Stamets, Chilton went on to be an industry leader in mushroom cultivation, founding Nammex Mushrooms, a giant in the mushroom extracts market.

49. Steinhardt, *Mycelium Is the Message*, 223.

50. Robert McPherson [Psylocybe Fanaticus], "The United States of America V.," accessed June 26, 2020, https://www.seanet.com/~rwmcpherson/pfcase .html#intro.

51. Letcher, *Shroom*, 279–80.

52. Maia Szalavitz, "Steve Jobs Had LSD. We Have the iPhone," *Time*, October 6, 2011, https://healthland.time.com/2011/10/06/jobs-had-lsd-we-have -the-iphone/.

53. Olivia Solon, "Under Pressure, Silicon Valley Workers Turn to LSD Microdosing," *Wired UK*, August 24, 2016, https://www.wired.co.uk/article /lsd-microdosing-drugs-silicon-valley.

54. Douglas Rushkoff, *Cyberia: Life in the Trenches of Hyperspace* (Manchester, UK: Clinamen Press, 1994), 49–50.

55. Douglas Rushkoff, "After Cyberia," *Team Human*, Virtual Futures Salon, video, 5:46, June 22, 2017, https://www.youtube.com/watch?v=jISdK4mW8gU.

56. "Glossary and Lexicon of the Online Mushroom Community's (OMC) terms," Shroomery.org (website), February 2, 2017, https://www.shroomery.org /13894/Glossary-and-Lexicon-of-The-Online-Mushroom-Communitys -OMC-terms.

57. McCoy, *Radical Mycology*, 241.

58. [Hippie3], "Airport re–deux," Mycotopia.net post, October 14, 2005, https:// mycotopia.net/topic/6079-airport-re-deux.

Chapter 6: Spawn Points

1. A. Elizabeth Sloan, "Top 10 Food Trends," *IFT News*, April 1, 2019, https:// www.ift.org/news-and-publications/food-technology-magazine/issues/2019 /april/features/2019-top-10-food-trends.

2. Laura Pitcher, "Indoor Neon-Lit Mushroom Farms Are New York's Hottest New Food Trend," *The Guardian*, November 1, 2018, https://www.theguardian

.com/food/2018/nov/01/indoor-neon-lit-mushroom-farms-are-new-yorks
-hottest-new-food-trend.

3. Ella Riley-Adams, "How This Brooklyn Artist Gets Dressed to Go Mushroom
Foraging," *Vogue*, February 24, 2020, https://www.vogue.com/vogueworld
/article/phyllis-ma-artist-mushrooms-zine-smallhold-brooklyn.

4. Nina Sparling, "How a Mushroom Farm Grows in a Manhattan Restaurant,"
Vogue, January 24, 2018, https://www.vogue.com/article/smallhold-mushroom
-urban-farming.

5. "Home Delivery Campaign—Strategy Guide and Toolkit," Smallhold (website),
April 1, 2020, https://docs.google.com/document/d/1JCdg0mj4cgO
_ygTOKJq_8B2SJgoI2rOBhkZHDtFMXIE/edit.

6. Andrew Carter, "When Sh*! Hits the Fan, Grow Mushrooms on It," LinkedIn
post, April 10, 2020, https://www.linkedin.com/pulse/when-sh-hits-fan-grow
-mushrooms-andrew-carter/.

7. Joseph N. DiStefano, "Kennett Square Mushroom Grower's $115 Million
Expansion Is a Big Bet for New York Investor," January 30, 2020, https://www
.inquirer.com/business/mushrooms-kennett-antitrust-champs-eos-pia-south
-mill-20200129.html.

8. Willis, ed., *State of the World's Fungi*, 51.

9. Peter Gillins, "Violence Clouds Wild Mushroom Harvest in U.S. as Demand
Takes Off," *Los Angeles Times*, August 1, 1993, https://www.latimes.com
/archives/la-xpm-1993-08-01-me-19088-story.html.

10. Defined as a "communal use and understanding of a specific interest and
knowledge of its best practices"; see "Communities of Interest, Communities
of Practice," *Encyclopedia.com*, August 4, 2020, https://www.encyclopedia.com
/management/encyclopedias-almanacs-transcripts-and-maps/communities
-interest-communities-practice.

11. In China, which has centuries of head start on North American cultivators,
morels and other tough-to-grow species are grown at scales that have yet to be
achieved here.

12. Cotter, *Organic Mushroom Farming*, 37–43.

13. Higher concentrations of alcohol are, seemingly paradoxically, less effective in
killing bacteria, evaporating faster, leaving less time in contact to break open
microbial cell walls.

14. "Antifungal Resistance," Centers for Disease Control and Prevention, National
Center for Emerging and Zoonotic Infectious Diseases, Division of Foodborne, Waterborne, and Environmental Diseases, May 18, 2020, https://www
.cdc.gov/fungal/antifungal-resistance.html.

15. Eugene Davenport, "Report of the Director of the Agricultural Experiment
Station University of Illinois," *University of Illinois* 33 (June 30, 1920), 50.

16. Cotter, *Organic Mushroom Farming*, 35.
17. Plastic is one of the persistent frustrations of sustainability-minded mushroom growers. Bags need to be capable of withstanding both sterilization at high temperatures, and the digestive process of a mycelium as it consumes its substrate.
18. James Hamblin, "The Mushrooms Are Slowly Taking Effect," *The Atlantic*, May 16, 2019, https://www.theatlantic.com/health/archive/2019/05/mushroom -law/589192/.
19. Alex Halperin, "What It's Like to Audition for 'Shark Tank' for Weed," *Rolling Stone*, June 1, 2016, https://www.rollingstone.com/politics/politics-news/what -its-like-to-audition-for-shark-tank-for-weed-46290/.
20. Cotter, *Organic Mushroom Farming*, 179–83.
21. Arevalo, *DIY Mushroom Cultivation*, 42.

Chapter 7: Building a Myco Scene

1. Radical Mycology, "Radical Mycology: Training a Mushroom to Remediate Cigarette Filters," video, 5:37, January 5, 2014, https://www.youtube.com /watch?v=fCAX9P50SNU.
2. "PSMS History and Mission," Puget Sound Mycological Society (website), accessed August 6, 2020, http://www.psms.org/history.php.
3. McCoy, *Radical Mycology*, vii.
4. McCoy, 379–401.
5. Joanna Steinhardt, "Psychedelic Naturalism and Interspecies Alliance: Views from the Emerging Do-It-Yourself Mycology Movement," *Plant Medicines, Healing and Psychedelic Science: Cultural Perspectives* (Cham, CH: Springer, 2018), 167–84, https://doi.org/10.1007/978-3-319-76720-8_10.
6. Edward Rothstein, "A Crunchy-Granola Path from Macramé and LSD to Wikipedia and Google," *New York Times*, September 25, 2006, https://www .nytimes.com/2006/09/25/arts/25conn.html.
7. McCoy, *Radical Mycology*, 380.
8. McCoy, 392.
9. "The Cascadia Bioregion: Facts & Figures," Cascadia Illahee Department of Bioregional Affairs (website), accessed August 7, 2020, https://deptofbioregion .org/facts-and-figures.
10. "II. Indians and Europeans on the Northwest Coast: Historical Context," Center for the Study of the Pacific Northwest, accessed June 26, 2020, https:// www.washington.edu/uwired/outreach/cspn/Website/Classroom%20Materials /Curriculum%20Packets/Indians%20&%20Europeans/II.html.
11. Al Burian, "A Night Out with an Anarchist," *Vice*, July 13, 2012, https://www .vice.com/en_us/article/dp4xbv/what-happened-to-crimethinc.
12. "Radical Mycology, an SLF Primer," 1st ed., Spore Liberation Front (2009).

13. "Hey Permaculture and Mycology Geeks the Radical Mycology Convergence Is Coming Up Soon!" *Punk Rock Permaculture E-Zine*, August 15, 2011, https://punkrockpermaculture.wordpress.com/2011/08/15/hey-permaculture-and-mycology-geeks-the-radical-mycology-convergence-is-coming-up-soon/.
14. McCoy, *Radical Mycology*, 388.
15. A fungus, identified in 2019, that metabolizes gold, suggests there may be little difference between "out there" and "ahead of the curve"; see Tsing Bohu et al., "Evidence for Fungi and Gold Redox Interaction Under Earth Surface Conditions," *Nature Communications* 10, no. 1 (May 23, 2019): 2290.
16. Peter Wohlleben, *The Hidden Life of Trees: What They Feel, How They Communicate—Discoveries from a Secret World* (London: HarperCollins Publishers, 2016), 53.
17. Steph Yin, "How Researchers and Advocates of Color Are Forging Their Own Paths in Psychedelic-Assisted Therapy," *WHYY*, January 10, 2020, https://whyy.org/segments/researchers-advocates-of-color-are-forging-their-own-paths-in-psychedelic-assisted-therapy/.
18. Olivia Goldhill, "A Millionaire Couple Is Threatening to Create a Magic Mushroom Monopoly," *Quartz*, November 8, 2018, https://qz.com/1454785/a-millionaire-couple-is-threatening-to-create-a-magic-mushroom-monopoly/.
19. *Deadnaming* is defined as "the use of a transgender person's birth-name without their consent."
20. Dr. Joanna Steinhardt, "Psychedelic Naturalism and Interspecies Alliance: Views from the Emerging Do-It-Yourself Mycology Movement," *Research Plant Medicines, Healing and Psychedelic Science: Cultural Perspectives* (2018): 169, https://doi.org/10.1007/978-3-319-76720-8_10.
21. Arevalo recalled the description offered by a mycological colleague to a shiitake-based dish he'd made: "It's as if I'm licking the armpit of the Buddha."
22. Anna Lowenhaupt Tsing, *The Mushroom at the End of the World: The Possibility of Life in Capitalist Ruins* (Princeton, NJ: Princeton University Press, 2015), 152.
23. Anna Lowenhaupt Tsing, "Unruly Edges: Mushrooms as Companion Species," *Environmental Humanities* 1, no. 1 (November 2012): 144–51.

Chapter 8: Ferment Yourself

1. Rich Shih and Jeremy Umansky, *Koji Alchemy: Rediscovering the Magic of Mold-Based Fermentation* (White River Junction, VT: Chelsea Green, 2020), 16.
2. Sandor Katz [sandorkraut], "The People's Republic of Fermentation," video series, July 1, 2017, https://www.youtube.com/playlist?list=PLDfUp9XK6kA176NN76_4vxx983PEGK9q_.
3. Jeff Gordinier, "Better Eating, Thanks to Bacteria," *New York Times*, September 17, 2012, https://www.nytimes.com/2012/09/19/dining/fermentation-guru-helps-chefs-find-new-flavors.html.

4. A. Elizabeth Sloan, "Favoring Fermented," *Institute of Food Technologists, Consumer Trends*, November 1, 2019, https://www.ift.org/news-and-publications/food-technology-magazine/issues/2019/november/columns/favoring-fermented.

5. Lizzy Saxe, "Fermented Foods Are Up 149%—As Long as They're Unfamiliar," *Forbes*, February 6, 2019, https://www.forbes.com/sites/lizzysaxe/2019/02/06/fermented-foods-are-up-149-percent-as-long-as-theyre-unfamiliar/.

6. Summary of "Global Fermented Food and Ingredients Market," BIS Research (March 2019), https://www.reportlinker.com/p05767424/Global-Fermented-Food-and-Ingredients-Market-Focus-on-Food-Type-Fermented-Dairy-Fermented-Beverages-Fermented-Vegetables-Ingredient-Type-Amino-Acid-Organic-Acid-Industrial-Enzyme-and-Distribution-Channel-Analysis-and-Forecast.html.

7. Katy Askew, "'There Is a Mega-Trend Around Fermentation': The Rising Star of Fermented Foods," *Food Navigator*, May 8, 2018, https://www.foodnavigator.com/Article/2018/05/04/There-is-a-mega-trend-around-fermentation-The-rising-star-of-fermented-foods.

8. Eater Staff, "Momofuku's Kaizen Trading Company Up and Fermenting," *Eater*, March 20, 2014, https://www.eater.com/2014/3/20/6258145/momofukus-kaizen-trading-company-up-fermenting.

9. René Redzepi and David Zilber, *The Noma Guide to Fermentation: Including Koji, Kombuchas, Shoyus, Misos, Vinegars, Garums, Lacto-ferments, and Black Fruits and Vegetables* (New York: Artisan, 2019), 13.

10. Sandor Ellix Katz, *Wild Fermentation: The Flavor, Nutrition, and Craft of Live-Culture Foods* (White River Junction, VT: Chelsea Green, 2016), 16–17.

11. Sandor Ellix Katz, *The Art of Fermentation: An In-Depth Exploration of Essential Concepts and Processes from Around the World* (White River Junction, VT: Chelsea Green, 2016), 114–16.

12. Sarah Zielinski, "The Alcoholics of the Animal World," *Smithsonian Magazine*, September 16, 2011, https://www.smithsonianmag.com/science-nature/the-alcoholics-of-the-animal-world-81007700/.

13. Attendance was not totally random, of course; Katz selects the group for each workshop from a large pool of applicants. Many had tried several times before finally being invited to attend.

14. Maria Prado et al., "Milk Kefir: Composition, Microbial Cultures, Biological Activities, and Related Products," *Frontiers in Microbiology* 6 (October 30, 2015): 1177, https://doi.org/10.3389/fmicb.2015.01177.

15. Silvia Alejandra Villarreal-Soto et al., "Understanding Kombucha Tea Fermentation: A Review," *Journal of Food Science* 83 (March 6, 2018): 580–88, https://doi.org/10.1111/1750-3841.14068.

16. Luba Vikhanski, "A Science Lecture Accidentally Sparked a Global Craze for Yogurt," *Smithsonian Magazine*, April 11, 2016, https://www.smithsonianmag

.com/science-nature/science-lecture-accidentally-sparked-global-craze
-yogurt-180958700/.

17. "One Yogurt, 100 Years of Science," Danone Blog, *Medium*, December 9, 2016, https://medium.com/@Danone/one-yogurt-100-years-of-science-34d353037cf4.

18. Alex Halberstadt, "Out of the Woods: After Decades of Semi-Secrecy, a Commune for L.G.B.T.Q. Nonconformists Has Slowly Begun to Join the Mainstream," *New York Times Magazine*, August 6, 2015, https://www.nytimes .com/2015/08/09/magazine/out-of-the-woods.html.

19. Halberstadt, "Out of the Woods."

20. Jennifer Hsu, "John, the Tempeh Man," WNYC (website), January 31, 2011, https://www.wnyc.org/story/112254-john-parker-bushwick-brooklyn -tempeh-man/.

21. Jim Windolf, "Sex, Drugs, and Soybeans," *Vanity Fair*, April 6, 2007, https:// www.vanityfair.com/news/2007/05/thefarm200705.

22. Albert K. Bates, "The Farm: Technological Innovation in a Rural Intentional Community, 1971–1987," The Farm, October 17, 1987, http://old.thefarm.org /lifestyle/albertbates/akbp1.html.

23. William Shurtleff and Akiko Aoyagi, *History of Tempeh and Tempeh Products (1815–2011), Extensively Annotated Bibliography and Sourcebook* (Lafayette, CA: Soyinfo Center, 2011), 583.

24. Shurtleff and Aoyagi, *History of Tempeh*, 415–16.

25. Andrew Macmillan (Partner, Farm Foods LLC), email messages to author, July 29, 2020.

26. The Brews Brothers, "What's Old Is New: Coolships in American Craft Brewing," *Craft Beer*, January 9, 2017, https://www.craftbeer.com/craft-beer -muses/coolships-old-new-american-craft-brewing.

27. Michael Tonsmeire, "Ambient-Spontaneous Yeast Starters," *The Mad Fermentationist*, April 25, 2011, https://www.themadfermentationist.com/2011/04 /ambient-spontaneous-yeast-starters.html.

28. Richard Preiss et al., "Traditional Norwegian Kveik Are a Genetically Distinct Group of Domesticated Saccharomyces Cerevisiae Brewing Yeasts," *Frontiers in Microbiology* 9 (September 12, 2018), https://doi.org/10.3389 /fmicb.2018.02137.

Chapter 9: Bio Prospects

1. Appropriately, *Cordyceps* was the inspiration for the fungal plague in the super popular zombie horror game, *The Last of Us*; see Kyle Hill, "The Fungus that Reduced Humanity to the Last of Us," *Scientific American*, June 25, 2013, https://blogs.scientificamerican.com/but-not-simpler/the-fungus-that -reduced-humanity-to-the-last-of-us/.

2. A stroma is essentially a mushroom: a fingerlike thallus with spore-bearing tissue covering its surface.

3. Douglas Rushkoff, *Throwing Rocks at the Google Bus: How Growth Became the Enemy of Prosperity* (New York: Penguin Random House, 2016), 16–24.

4. Ruma Arora Soni et al., "Spirulina: From Growth to Nutritional Product: A Review," *Trends in Food Science & Technology* 69, part A (September 2017): 157–71, https://doi.org/10.1016/j.tifs.2017.09.010.

5. Gi-Ho Sung et al., "Phylogenetic Classification of Cordyceps and the Clavicipitaceous Fungi," *Studies in Mycology* 57 (2007): 5–59, https://dx.doi.org/10.3114%2Fsim.2007.57.01.

6. In 2016, a former Chinese police official in Guangdong Province was sentenced to fifteen years of prison after a stash of some 440 pounds of the "Himalayan Viagra" was discovered in his home, worth an estimated $9.4 million; see Lucy Papachristou, "China Cracks Down on 'Himalayan Viagra' Used to Bribe Officials," *Organized Crime and Corruption Reporting Project*, March 25, 2019, https://www.occrp.org/en/daily/9443-china-cracks-down-on-himalayan-viagra-used-to-bribe-officials.

7. Report summary "Mushroom Market Size, Share & Trends Analysis Report by Application, Regional Outlook, Competitive Strategies, and Segment Forecasts, 2019 to 2025," *Grand View Research*, accessed June 28, 2020, https://www.grandviewresearch.com/industry-analysis/mushroom-market.

8. Hardeep S. Tuli et al., "Pharmacological and Therapeutic Potential of Cordyceps with Special Reference to Cordycepin," *3 Biotech* 4, no. 1 (February 2014): 1–12, https://dx.doi.org/10.1007%2Fs13205-013-0121-9.

9. Bai-Xiong Chen et al., "Transcriptome Analysis Reveals the Flexibility of Cordycepin Network in Cordyceps Militaris Activated by L-Alanine Addition," *Frontiers in Microbiology* 11 (April 24, 2020): 577, https://doi.org/10.3389/fmicb.2020.00577.

10. Rina Raphael, "The 'Shroom Boom': Will Trendy Medicinal Mushrooms Go Mainstream in 2018?" *Fast Company*, January 6, 2018, https://www.fastcompany.com/40511575/the-shroom-boom-will-trendy-medicinal-mushrooms-go-mainstream-in-2018.

11. Willy Crosby, *Cordyceps Mushroom Cultivation in the Northeast United States* (College Park, MD: USDA Sustainable Agriculture Research and Education Program, 2019), https://cdn.sare.org/wp-content/uploads/20190327213115/FungiAlly_Cordycepts_booklet_FINAL.pdf.

12. *Cordyceps* are part of the phylum Ascomycota, and the threadlike spores they release are called ascospores.

13. Alex Dorr and Lera Niemackl, interview with Ryan Paul Gates, *Mushroom Revival Podcast*, podcast audio, June 25, 2019, https://www.mushroom-revival.com/reishi-and-cordyceps-genetics-with-ryan-paul-gates-episode-9/.

14. Cotter, *Organic Mushroom Farming*, 148–49.
15. Ants infected with *Cordyceps* spores are systematically removed by their nest mates; see Carl Zimmer, "After This Fungus Turns Ants into Zombies, Their Bodies Explode," *New York Times*, October 24, 2019, https://www.nytimes .com/2019/10/24/science/ant-zombies-fungus.html.
16. Stamets, *Mycelium Running*, 58.
17. Cotter, *Organic Mushroom Farming*, 205–208.

Chapter 10: Applied Mycology

1. Russell T. Graham, ed., *Hayman Fire Case Study, General Technical Report, RMRS-GTR-114* (Ogden, UT: USDA, Forest Service, Rocky Mountain Research Station, 2018), 28, https://www.fs.fed.us/rm/pubs/rmrs _gtr114.pdf.
2. Ten years after Hayman, almost to the day, the Waldo Canyon Fire ignited west of Colorado Springs, killing two people, leveling 346 homes, and devastating some eighteen thousand acres. Six years after that, the Spring Fire burned more than 108,000 acres, Colorado's second-largest fire. Five of the twenty largest wildfires in the state's history took place in 2018; see Kirk Mitchell, "Updated for 2018: 20 Largest Wildfires in Colorado History by Acreage Burned," *Denver Post*, July 4, 2018, https://www.denverpost.com/2018 /07/04/20-largest-wildfires-in-colorado-history-in-acreage-burned/.
3. Ellie Mulder, "Woman Who Started Largest Fire in Colorado History Sentenced to 15 More Years of Probation," *Gazette* (Colorado Springs, CO), August 16, 2018, https://gazette.com/news/woman-who-started-largest -fire-in-colorado-history-sentenced-to/article_195b99ea-a19e-11e8-8619 -3f2fc4d12ea3.html.
4. Daniel A. Yaussy, ed., *Proceedings: Workshop on Fire, People, and the Central Hardwoods Landscape, March 12–14, 2000, Richmond, Kentucky* (Newtown Square, PA: USDA Forest Service, 2000), 19–27, https://www.fs.fed.us/ne/ newtown_square/publications/technical_reports/pdfs/2000/gtrne274.pdf.
5. "History of the Colorado State Forest," Colorado State Forest Service, accessed June 28, 2020, https://csfs.colostate.edu/colorado-state-forest/csf-history/.
6. *2016 Report on the Health of Colorado's Forests: Fire and Water* (Fort Collins, CO: Colorado State Forest Service, January 2017), https://csfs.colostate.edu/media /sites/22/2017/02/CSU_304464_ForestReport-2016-www.pdf.
7. Scott Gondon, "Forest Recovery from Lake Christine Fire Could Take Centuries," *Aspen Times*, January 13, 2019, https://www.denverpost.com/2019/01/13 /lake-christine-wildfire-recovery/.
8. "Wildfire Recovery & Flood Mitigation," Coalition for the Upper South Platte, accessed June 28, 2020, https://cusp.ws/flood-mitigation/.

9. There are other colors, too. Some beetle-borne fungi leave the wood they eat a shade of blue, green, or black, a result of the mycelium's own color. Lumber companies often sell the spalted wood as a value-added product.

10. Moore, *Fungal Biology*, 9–11.

11. Colorado Mycological Society, "Jeff Ravage Mycoremediation," (June 8, 2020), Facebook Live seminar, https://www.facebook.com/andy.wilson.9279807 /videos/10158651820008701/.

12. Ravage insists the rescued bird had fewer feathers when he first found it.

13. Lauren M. Czaplicki et al., "A New Perspective on Sustainable Soil Remediation—Case Study Suggests Novel Fungal Genera Could Facilitate in situ Biodegradation of Hazardous Contaminants," *Remediation* 26, no. 2 (March 2, 2016): 59–72, https://dx.doi.org/10.1002%2Frem.21458.

14. Robert A. Zabel et al., "Fungi Associated with Decay in Treated Southern Pine Utility Poles in the Eastern United States," *Wood and Fiber Science* 17, vol. 1 (1985): 75–91.

15. Kelechi Njoku et al., "Mycoremediation of Petroleum Hydrocarbon Polluted Soil by Pleurotus Pulmonarius," *Ethiopian Journal of Environmental Studies and Management* 9, no. 1 (2016), https://doi.org/10.4314/ejesm.v9i1.6s.

16. Alex Taylor et al., "Removal of *Escherichia coli* from Synthetic Stormwater Using Mycofiltration," *Ecological Engineering* 78 (May 2015): 79–86, https:// doi.org/10.1016/j.ecoleng.2014.05.016.

17. Czaplicki et al., "Sustainable Soil Remediation," 59–72.

18. "Superfund Site: Atlantic Wood Industries, Inc. Portsmouth, VA," Environmental Protection Agency, accessed June 28, 2020, https://cumulis.epa.gov/supercpad /SiteProfiles/index.cfm?fuseaction=second.Cleanup&id=0302836#bkground.

19. "Great Lakes AOCs: Maumee AOC," Environmental Protection Agency, accessed August 8, 2020, https://www.epa.gov/great-lakes-aocs/maumee-aoc.

20. Jane Kay and Cheryl Katz, "Pollution, Poverty and People of Color: Living with Industry," *Scientific American*, June 4, 2012, https://www.scientificamerican .com/article/pollution-poverty-people-color-living-industry/.

21. "Manifesto," *The Pueblo/a/x Feminist Caucus*, The Red Nation (website), 2019, https://therednationdotorg.files.wordpress.com/2019/09/pfc-manifesto.pdf.

22. Christine Darosa, "The Pueblo Revolt of 1680," *Socialist Worker*, August 17, 2011, https://socialistworker.org/2011/08/17/the-pueblo-revolt-of-1680.

23. Maya Salam, "Native American Women Are Facing a Crisis," *New York Times*, April 12, 2019, https://www.nytimes.com/2019/04/12/us/native-american -women-violence.html.

24. Executive summary to "Toxic Waste and Race, in the United States: A National Report on the Racial and Socio-Economic Characteristics of Communities with Hazardous Waste Sites" (New York: United Church of Christ

Commission for Racial Justice, 1987), xiv, https://www.nrc.gov/docs/ML1310
/ML13109A339.pdf.

25. *Cancer Among American Indians in New Mexico 2008–2013* (Santa Fe: New
Mexico Department of Health, 2014), 4, https://www.nmhealth.org/data
/view/chronic/1788/.

26. "Potential Release Sites," Los Alamos National Laboratory, accessed June 28,
2020, https://www.lanl.gov/community-environment/environmental
-stewardship/cleanup/sites/potential-release-sites.php.

27. Joseph Masco, *The Nuclear Borderlands: The Manhattan Project in Post–Cold War
New Mexico* (Princeton, NJ: Princeton University Press, 2006), 149.

28. Staci Matlock, "'Los Alamos Will Never Be Clean,'" *Santa Fe New Mexican*, July
13, 2005, https://www.santafenewmexican.com/news/local_news/los-alamos
-will-never-be-clean/article_a3cc7ce1-8af0-5113-8f38-5d4aa673fd7a.html.

29. John Fleck, "Open Burning Ban at LANL Reconsidered," *Albuquerque Journal*,
January 4, 2011, abqjournal.com/news/state/04225952state01-04-11.htm.

30. N3B, "Los Alamos National Laboratory RDX Characterization Fact Sheet,"
Los Alamos National Laboratory (website), February 2019, https://www
.energy.gov/sites/prod/files/2019/03/f61/RDX%20Fact%20Sheet_February
%202019_FINAL.pdf.

31. Eileen Welsome, *The Plutonium Files* (New York: Dell Publishing, 1999), 70–71.

32. Rebecca Moss, "Half Life," *ProPublica*, October 26, 2018, https://features
.propublica.org/los-alamos/chad-walde-nuclear-facility-radiation-cancer/.

33. Morgan Drewniany, "Red Dust: A Soil Scientist's Journey Through the Political
Climate and Environmental Chemistry of Northern New Mexico," (Division
III project, Hampshire College, 2013), 128–29, https://compass.fivecolleges
.edu/object/hampshire:1526.

34. N3B, "Los Alamos National Laboratory Chromium Plume Fact Sheet," Los
Alamos National Laboratory (website), August 28, 2018, https://www.energy
.gov/sites/prod/files/2018/11/f57/Chromium%20Plume%20Fact%20Sheet
%20Fall%202018.pdf.

35. Max Genecov, "Still Toxic After All These Years," *Grist*, January 29, 2019,
https://grist.org/article/the-true-story-of-the-town-behind-erin-brockovich/.

36. The EPA defines a sole-source aquifer as one that supplies at least 50 percent
of the drinking water for its service area, with no reasonable alternative should
the aquifer become contaminated; see "Overview of the Drinking Water Sole
Source Aquifer Program," EPA, accessed June 28, 2020, https://www.epa.gov
/dwssa/overview-drinking-water-sole-source-aquifer-program.

37. Another, unrelated groundwater plume of trichloroethylene (TCE) and other
chemicals underneath a dry cleaner's in Española resulted in its designation as
a federal Superfund site; see "Superfund Site: North Railroad Avenue Plume,

Espanola, NM," Environmental Protection Agency, accessed August 8, 2020, https://cumulis.epa.gov/supercpad/SiteProfiles/index.cfm?fuseaction=second .Cleanup&id=0604299#bkground.

38. "LANL Updates Public on Chromium Contamination," Citizens Concerned for Nuclear Safety, accessed August 8, 2020, http://www.nuclearactive.org /news/041307.html.

39. Office of NEPA and Policy Compliance, "EA-2005: Chromium Plume Control Interim Measure And Plume-Center Characterization, Los Alamos National Laboratory, Los Alamos, NM," US Department of Energy, accessed August 8, 2020, https://www.energy.gov/nepa/ea-2005-chromium-plume-control -interim-measure-and-plume-center-characterization-los-alamos.

40. Mark Oswald, "High Chromium Levels Found at One Los Alamos Well," *Albuquerque Journal*, September 18, 2017, https://www.abqjournal.com/1065665 /high-chromium-readings-found-beyond-los-alamos-plume-boundary.html.

41. Kendra Chamberlain, "A Long Road to Remediation for Hexavalent Chromium Plume Near Los Alamos," *NM Political Report*, August 26, 2019, https://nmpoliticalreport.com/2019/08/26/a-long-road-to-remediation -for-hexavalent-chromium-plume-near-los-alamos/.

42. DOE/LANS, "Multiple Activities Work Plan for the Treatment and Land Application of Groundwater from Mortandad and Sandia Canyons, DP–1793," NMED Ground Water Quality Bureau, Permit Enclosure 1 (March 22, 2016), https://permalink.lanl.gov/object/tr?what=info:lanl-repo /eprr/ESHID-601362.

43. "In the Matter of Los Alamos National Laboratory's Groundwater Discharge Permit, DP–1793," Before the Office of the Secretary, Transcript of Proceedings, vol. 1, (Los Alamos, NM: State of New Mexico, November 7, 2018).

44. Deonna Anderson, "Mushrooms: 'Nature's Greatest Decomposers,'" *Yes!*, April 16, 2019, https://www.yesmagazine.org/issue/dirt/2019/04/16/food-natural -way-clean-toxic-soil-mushrooms/.

45. Parminder Chahal, "Glyphosate-Resistant Palmer Amaranth Confirmed in South-Central Nebraska," *Cropwatch*, March 15, 2017, https://cropwatch.unl.edu /2017/glyphosate-resistant-palmer-amaranth-confirmed-south-central-nebraska.

Chapter 11: Myx Messages

1. Steven Johnson, *Emergence: The Connected Lives of Ants, Brains, Cities, and Software* (London: Penguin Books, 2001), 18.

2. In the fall of 2019, the Paris Zoo even added a *Physarum polycephalum* to its list of attractions, under the name, "Le Blob"; see Thierry Chiarello and Kathryn Carlson, "Paris Zoo Unveils the 'Blob,' an Organism with No Brain But 720 Sexes," *Reuters*, October 16, 2019, https://www.reuters.com/article

/us-france-zoo-blob/paris-zoo-unveils-the-blob-an-organism-with-no-brain
-but-720-sexes-idUSKBN1WV2AD.

3. Cyd Cipolla et al., eds., *Queer Feminist Science Studies: A Reader* (Seattle: University of Washington Press, 2017), 310–25.

4. Romain P. Boisseau, "Habituation in Non-Neural Organisms: Evidence from Slime Moulds," *Proceeding of Royal Sciences B* (April 27, 2016), 28, https:// royalsocietypublishing.org/doi/10.1098/rspb.2016.0446.

5. David Vogel and Audrey Dussutour, "Direct Transfer of Learned Behaviour Via Cell Fusion in Non-Neural Organisms," *Proceeding of Royal Sciences B* (December 28, 2016), 28, https://royalsocietypublishing.org/doi/10.1098/rspb.2016.2382.

6. Andrew Adamatzky, "Slime Mold Solves Maze in One Pass, Assisted by Gradient of Chemo-Attractants," *IEEE Transactions on Nanobioscience* 11 (2012): 131–34, https://doi.org/10.1109/TNB.2011.2181978.

7. Laura Sanders, "Slime Mold Grows Network Just Like Tokyo Rail System," *Wired*, January 22, 2010, https://www.wired.com/2010/01/slime-mold-grows -network-just-like-tokyo-rail-system/.

8. Franklin Patterson and Charles R. Longsworth, "Hampshire College as an Instrument of Change," in *Making of a College*, special fall 2004 ed. (Cambridge, MA: MIT Press, 1975), 23–24.

9. Andrew Adamatzky, ed., *Slime Mould in Arts and Architecture* (Alsbjergvej, DK: River Publishers, 2019), 61–88.

10. Albert Goldbeter, "Mechanism for Oscillatory Synthesis of Cyclic AMP in Dictyos- telium Discoideum," *Nature* 253 (1975): 540–42, https://doi.org/10.1038/253540a0.

11. Carlos H. Serezani et al., "Cyclic AMP: Master Regulator of Innate Immune Cell Function," *American Journal of Respiratory Cell and Molecular Biology* 39, no. 2 (August 2008): 127–32, https://dx.doi.org/10.1165%2Frcmb.2008-0091TR.

12. Birgit M. Shaffer, "The Acrasina," *Advances in Morphogenesis* 2 (1962): 109–82, https://doi.org/10.1016/B978-1-4831-9949-8.50007-9.

13. Johnson, *Emergence*, 14–15.

14. Evelyn F. Keller and Lee A. Segel, "Initiation of Slime Mold Aggregation Viewed as an Instability," *Journal of Theoretic Biology* 26, no. 3 (March 1970): 399–415, https://doi.org/10.1016/0022-5193(70)90092-5.

15. Alan Mathison Turing, "The Chemical Basis of Morphogenesis," *Philosophical Transactions of the Royal Society of London B*. 237, no. 641 (August 14, 1952): 37–72, https://doi.org/10.1098/rstb.1952.0012.

16. Sherry Turkle, ed., *Evocative Objects: Things We Think With* (Cambridge, MA: MIT Press, 1985), 296–306.

17. Evelyn Fox Keller, "The Force of the Pacemaker Concept in Theories of Aggregation in Cellular Slime Mold," *Perspectives in Biology and Medicine* 26, no. 4 (Summer 1983): 515–21, https://doi.org/10.1353/pbm.1983.0049.

18. David Moore et al., "Laboulbeniales," in *21st Century Guidebook to Fungi*, 2nd ed., http://www.davidmoore.org.uk/21st_Century_Guidebook_to_Fungi _PLATINUM/Ch16_04.htm.

19. Alex Weir, Gordon Beakes, "An Introduction to the Laboubeniales: A Fascinating Group of Entomogenous Fungi," *Mycologist* 9, no. 1 (February 1995): 6, https://doi.org/10.1016/S0269-915X(09)80238-3.

20. Patricia Kaishian and Hasmik Djoulakian, "The Science Underground: Mycology as a Queer Discipline," *Catalyst: Feminism, Theory and Science* 6, no. 2 (Nov 7, 2020), https://doi.org/10.28968/cftt.v6i2.33523.

21. David L. Hawksworth et al., "Mycology: A Neglected Megascience," *Applied Mycology* (2009): 1–17, https://doi.org/10.1079/9781845935344.0001.

22. Robert Scott and Stephen Kosslyn, eds., *Emerging Trends in the Social and Behavioral Sciences* (Hoboken, NJ: John Wiley & Sons, 2015), 1–16.

23. Robin Wall Kimmerer, *Braiding Sweetgrass: Indigenous Wisdom, Scientific Knowledge, and the Teachings of Plants* (Minneapolis, MN: Milkweed Editions, 2013), 49.

24. Kimmerer, *Braiding Sweetgrass*, 55.

25. William Delisle Hay, *An Elementary Textbook of British Fungi* (London: Swan Sonnenschein, Lowrey & Co, 1887), 6–7.

26. Arthur Conan Doyle, "Sir Nigel," *Strand Magazine* 180, December 1905, 604.

27. Catriona Mortimer-Sandilands and Bruce Erickson, introduction to *Queer Ecologies: Sex, Politics, Nature, Desire* (Bloomington, Indiana State Press, 2010), 2–3.

28. Andrew Liptak, "Rivers Solomon on Colonialism, the Apocalypse, and Fascinating Fungus," *The Verge*, January 28, 2019, https://www.theverge.com /2019/1/28/18196370/rivers-solomon-interview-sci-fi-colonialism-fungus -better-world.

29. Donna J. Haraway, *When Species Meet* (Minneapolis: University of Minnesota Press, 2008), 19.

30. Tanya E. Cheeke et al., "Diversity in the Mycological Society of America," accessed August 8, 2020, https://msafungi.org/wp-content/uploads/2019/04/ inoculum-march-2018-Diversity-in-the-Mycological-Society-of-America.pdf.

31. Ed Yong, "When Will the Gender Gap in Science Disappear?" *The Atlantic*, April 19, 2018, https://www.theatlantic.com/science/archive/2018/04/when -will-the-gender-gap-in-science-disappear/558413/.

32. Merrill Perlman, "How the Word 'Queer' Was Adopted by the LGBTQ Community," *Columbia Journalism Review*, January 22, 2019, https://www.cjr .org/language_corner/queer.php.

Chapter 12: Who Speaks for the Mushroom

1. "Campaña Anti-Mecheros," UDAPT, accessed June 28, 2020, http://www.udapt .org/en/campana-anti-mecheros-2/.

2. Michael Cepek, "A Trip Through Ecuador's Cofán Community—and Its Disappearing Homeland," *Pacific Standard*, September 23, 2018, https://psmag.com/magazine/a-trip-through-ecuadors-cofan-community-and-its-disappearing-homeland.

3. Aoghs.org eds., "Sour Lake Produces Texaco," *American Oil & Gas Historical Society*, April 18, 2020, https://aoghs.org/petroleum-pioneers/sour-lake-produces-texaco.

4. Richard Ward, "In the Oil Fields of Ecuador," *Counterpunch*, April 18, 2014, https://www.counterpunch.org/2014/04/18/in-the-oil-fields-of-ecuador/.

5. Steven Donziger, "Chevron's 'Amazon Chernobyl' in Ecuador: The Real Irrefutable Truths About the Company's Toxic Dumping and Fraud," *Huffington Post*, May 27, 2016, https://www.huffpost.com/entry/chevrons-amazon-chernobyl_b_7435926.

6. Simon Romero and Clifford Krauss, "In Ecuador, Resentment of an Oil Company Oozes," *New York Times*, May 14, 2009, https://www.nytimes.com/2009/05/15/business/global/15chevron.html.

7. "Chevron Wins Ecuador Rainforest 'Oil Dumping' Case," *BBC News*, September 8, 2018, https://www.bbc.com/news/world-latin-america-45455984.

8. Maddie Stone, "The Plan to Mop Up the World's Largest Oil Spill with Fungus," *Vice Motherboard*, March 5, 2015, https://www.vice.com/en_us/article/jp5k9x/the-plan-to-mop-up-the-worlds-largest-oil-spill-with-fungus.

9. Incidentally, CoRenewal is also working on deploying mycorrhizal fungi and inoculated swales to foster post-fire recovery in California. Among those involved is Danielle Stevenson at UC Riverside, the mycorrhizal researcher from chapter 2.

10. G. Jock Churchman and Edward R. Landa, eds., *The Soil Underfoot* (Boca Raton, FL: CRC Press, 2014), 235–44.

11. Churchman and Landa, eds., *The Soil Underfoot*, 240.

12. Bruno Glaser, "Prehistorically Modified Soils of Central Amazonia: a Model for Sustainable Agriculture in the Twenty-First Century," in *Philosophical Transactions of the Royal Society of London, Series B, Biological Sciences* 362, no. 1478 (2007): 187–96.

13. James Fraser et al., "Anthropogenic Soils in the Central Amazon: From Categories to a Continuum," *Royal Geographical Society* 43, no. 3 (September 2011): 264–73, https://doi.org/10.1111/j.1475-4762.2011.00999.x.

14. Fraser et al., "Anthropogenic Soils," 265.

15. A. Keliikuli et al, "Natural Farming: The Development of Indigenous Microorganisms Using Korean Natural Farming Methods," in *Sustainable Agriculture*, College of Tropical Agriculture and Human Resources, University of Hawai'i at Mānoa (February 2019), https://www.ctahr.hawaii.edu/oc/freepubs/pdf/SA-19.pdf.

16. Not to be confused with the tek for cultivating *Psilocybe* mushrooms.

17. "Who We Are," UDAPT, accessed June 6, 2020, http://texacotoxico.net/en /who-we-are/.

18. Elaine R. Ingham, "Soil Food Web," USDA Natural Resources Conservation Service, accessed July 25, 2020, https://www.nrcs.usda.gov/wps/portal/nrcs /detailfull/soils/health/biology/?cid=nrcs142p2_053868.

19. Gabriela Coronel Vargas et al., "Public Health Issues from Crude-Oil Production, in the Ecuadorian Amazon territories," *Science of the Total Environment*, 719, no. 1 (June 1, 2020), https://doi.org/10.1016/j.scitotenv.2019.134647.

20. Anna-Karin Hurtig and Miguel San Sebastián, "Geographical Differences in Cancer Incidence in the Amazon Basin of Ecuador in Relation to Residence Near Oil Fields," *International Journal of Epidemiology* 31, no. 5 (October 2002): 1021–27, https://doi.org/10.1093/ije/31.5.1021.

21. Seema Patel and Arun Goya, "Recent Developments in Mushrooms as Anti-Cancer Therapeutics: A Review," *3 Biotech* 2 (November 25, 2011): 1–15, https://dx.doi.org/10.1007%2Fs13205-011-0036-2.

22. Robert Reid, "Explore the Cultural Heart of San Diego," *National Geographic*, accessed August 8, 2020, https://www.nationalgeographic.com/travel/destinations /north-america/united-states/california/san-diego/balboa-park-what-to-do/.

23. A form of corn smut, huitlacoche appears as silver gray bulbs that burst from the corn kernels; it's a savory, traditional South and Central American delicacy sometimes called Mexican truffle, or more humorously, "porn on the cob."

24. Kaitlin Sullivan, "Black Americans Are Being Left Out of Psychedelics Research," *Vice*, October 1, 2018, https://www.vice.com/en_us/article/pa834m /black-americans-are-being-left-out-of-psychedelics-research.

25. Carmen Chandler, "Only White People Can Get Away with the Microdosing Trend," *Vice*, December 19, 2018, https://www.vice.com/en_us/article/d3bj8w /only-white-people-can-get-away-with-the-microdosing-trend.

26. Steinhardt, "Mycelium Is the Message," 381–84.

27. Lucy Kavaler, *Mushrooms, Molds and Miracles* (New York: The John Day Company, 1965), 85–99.

28. Christina Troitino, "Is Meal Replacement Startup Soylent Down to Its Last Drop?" *Forbes* (February 18, 2020), https://www.forbes.com/sites /christinatroitino/2020/02/18/is-soylent-down-to-its-last-drop/.

29. Douglas Rushkoff, *Throwing Rocks*, 82–92.

30. Matthew Gault, "The World Economic Forum Tells Davos: Electronics Are 'the Fastest-Growing Waste Stream in the World,'" *Vice*, January 29, 2019, https://www.vice.com/en_us/article/8xynba/world-economic-forum-at-davos -electronics-are-the-fastest-growing-waste-stream-in-the-world.

31. Dan Chodorkoff, *The Anthropology of Utopia: Essays on Social Ecology and Community Development* (Grenmarsvegen, NO: New Compass Press, 2014), 145–58.

Index

INDEX

INDEX

INDEX

INDEX

United Kingdom (UK)
  Darwin Tree of Life Project, 69
  development of mycology, 70–74
  fungiphobia term, 251
  introduction of wood rot fungus, 158
  Kew Royal Botanic Gardens, 63–70, 71–72
  LAFF (Lost and Found Fungi project), 64–70
  mycological societies, 67, 71–72
  Red List, 65, 66
United States (US)
  fungi documentation project, 53–54
  hippie movement, 98
  mycological societies, 101
  online communities of mycophiles, 74
  See also North America; *specific states*
University of Utah
  foray into Uinta Mountains, 45–46
  fungarium, 61–62
  genetics laboratory, 49–54
  laboratory examination of finds, 46–47
  studies of Great Salt lake microbiology, 60
Utah
  bark beetle infestations, 39
  Dentinger, Bryn fungi documentation
    project, 53–54
  Great Salt Lake fungi, 54–61
Ute people, presence at Great Salt Lake, 57–58

Vancouver Mycological Society, 99, 100,
    101–2, 132
Veggie Mamacitx, 270
Velada ritual, 95

Wakefield, Elsie, 73
Wallace, Robert, 265
Wasson, R. Gordon
  Life magazine article, 95, 104
  Soma representation, 24
  studies of sacred mushroom practices,
    95–96, 97, 98, 105, 151

water contamination
  from Los Alamos National Laboratories,
    230–33
  mycofiltration for, 205
  from oil production, 256–57, 264
Weir, Alex, 247
White Labs, 176
white rot fungi, 213, 219, 220–21
Whitman, Mercedes Perez, 127
wild fermentation, 177–78, 179
Wild Fermentation (Katz), 171, 179
Wild Mushroom Food Safety Certification
  Course, 207
Wineka, Sam, 178
Wingfield, Michael J., 41
Wingo, Kai, 269
wood-rotting fungi, 158, 212–19
wood-wide-web, as term, 37
Woolhope Naturalists' Field Club (UK),
    70–71
World Trade Organization protests
    (1999), 143
Wright, Rich, 73, 74
Wyeast, 176

yeasts
  beer brewing, 175–180
  Bootleg Biology specialty yeast strains,
    181–82
  wild, 175, 177–78, 179–180
Yellowstone National Park, mountain pine
  beetle infestations, 39–40
yogurt, cultures in, 165–66
Yonah Schimmel, 166
Young, Edward, 58

Zalar, Polona, 56
Zilber, David, 162
ZoBell, Claude, 60
zygomycetes, 20

About the Author

ALANNA BURNS

Doug Bierend is a freelance journalist writing about science and technology, food, and education, and the various ways they point to a more equitable and sustainable world. His byline appears in *Wired*, *The Atlantic*, *Vice*, *Motherboard*, *The Counter*, *Outside*, *Civil Eats*, and numerous other publications.